AF443327

Designing and Developing Scalable IP Networks

Designing and Developing Scalable IP Networks

Guy Davies
Telindus, UK

John Wiley & Sons, Ltd

Published 2004 by   John Wiley & Sons Ltd, The Atrium, Southern Gate, Chichester,
West Sussex PO19 8SQ, England

Telephone (+44) 1243 779777

Email (for orders and customer service enquiries): cs-books@wiley.co.uk
Visit our Home Page on www.wileyeurope.com or www.wiley.com

This publication is designed to provide accurate and authoritative information in regard to the subject matter covered. It is sold on the understanding that the Publisher is not engaged in rendering professional services. If professional advice or other expert assistance is required, the services of a competent professional should be sought.

Other Wiley Editorial Offices

John Wiley & Sons Inc., 111 River Street, Hoboken, NJ 07030, USA

Jossey-Bass, 989 Market Street, San Francisco, CA 94103-1741, USA

Wiley-VCH Verlag GmbH, Boschstr. 12, D-69469 Weinheim, Germany

John Wiley & Sons Australia Ltd, 33 Park Road, Milton, Queensland 4064, Australia

John Wiley & Sons (Asia) Pte Ltd, 2 Clementi Loop #02-01, Jin Xing Distripark, Singapore 129809

John Wiley & Sons Canada Ltd, 22 Worcester Road, Etobicoke, Ontario, Canada M9W 1L1

Wiley also publishes its books in a variety of electronic formats. Some content that appears in print may not be available in electronic books.

Library of Congress Cataloging-in-Publication Data

Davies, Guy.
  Designing & developing scalable IP networks / Guy Davies.
      p. cm.
  Includes bibliographical references and index.
  ISBN 0-470-86739-6 (cloth : alk. paper)
  1. Computer networks—Design and construction. 2. Computer networks—Scalability. I. Title: Designing and developing scalable IP networks. II. Title.
  TK5105.5.D3794 2004
  004.6′2—dc22

                            2004011563

British Library Cataloguing in Publication Data

A catalogue record for this book is available from the British Library

ISBN 0-470-86739-6

Typeset in 10/12pt Times by Laserwords Private Limited, Chennai, India
Printed and bound in Great Britain by Antony Rowe Ltd, Chippenham, Wiltshire
This book is printed on acid-free paper responsibly manufactured from sustainable forestry in which at least two trees are planted for each one used for paper production.

Contents

List of Figures

List of Tables

About the Author

Guy Davies has worked as an IP Architect for Telindus, a network integrator in the UK, for four years. While at Telindus, he has been involved in many projects ranging from the design, implementation and operation of customers' core IP networks through to migration of ATM networks to an IP and MPLS-based infrastructure, and the design and implementation of large secured wireless networks and consultancy as well as the development of wireless rural broadband access. During this time, he has also worked as a contractor to Juniper Networks, providing engineering and consulting services both in the UK and overseas.

Prior to Telindus, Guy spent five years working for UUNET (and its previous incarnation in the UK, PIPEX). During his time at PIPEX and UUNET, Guy worked in a variety of engineering and management roles including systems administration, network operation and development roles. He was responsible for the design of the first pan-European MPLS core network built by UUNET.

Guy is JNCIE #20 and is also a CCIP.

BIOGRAPHY OF REVIEWER

Todd M. Regonini has almost ten years of industry experience in IP network design, implementation, and administration. He is currently employed as a Senior Systems Engineer working for Juniper Networks in Sunnyvale, California. Prior to that he worked for Cisco Systems where he consulted directly with customers on the design and implementation of IP networks. Todd is Juniper Networks Certified Internet Expert (JNCIE) #23.

Acknowledgements

This book has just one author. Despite that, it is only through the efforts of a number of people that this book has now been published.

I would first of all like to thank Patrick Ames. It was he who convinced me to write this book in the first place. Then, having got me to put my ideas down, it was he who helped me find a publisher for my work and he who kept me going, even when I doubted my own ability to complete the project. Without a doubt, this book would not have seen the light of day but for him.

Throughout the writing process, I have received corrections and advice from an array of friends and colleagues. Several engineers and consultants at Juniper Networks provided significant input. Particular thanks go to Aviva Garrett, Paul Goyette, Pete Moyer, Bill Nowak and Todd Regonini, who all provided invaluable technical information. I thank my former colleagues at UUNET, Tony Barber and John Whyte, for their thorough technical review and discussion of various issues covered in the book. I'd also like to thank Danny McPherson, who provided some useful insight into several issues surrounding BGP and the selection of parameters for policy. In addition to the technical reviewers above, Cressi Whyte gave some extremely useful advice on writing style.

I also received great help from Richard Southern of Juniper TAC in checking the configurations at the end of this book. I greatly appreciate the help he gave me at exceptionally short notice. Despite the time constraints, his work was thorough and accurate.

I would also like to thank the editing team at John Wiley & Sons, Ltd: Sally Mortimore, Birgit Gruber, Daniel Gill, Susan Dunsmore and Lucy Bryan.

Finally, I would especially like to thank my wife, Ghillie, and my children, Emily and George, for their endless patience. This book has taken much of my free time and most evenings for several months. I am sorry I have not paid as much attention to you or spent as much time with you as I should have done.

Abbreviations

AAA	Authorization, Authentication and Accounting
AAL	ATM Adaption Layer
ABR	Area Border Router
ABR	Available Bit Rate
AES	Advanced Encryption Standard
AH	Authentication Headers
APS	Automatic Protection Switching
ARP	Address Resolution Protocol
AS	Autonomous System
ASIC	Application Specific Integrated Circuit
ASM	Any Source Multicast
ATM	Asynchronous Transfer Mode
BA	Behaviour Aggregate
BECN	Backward Explicit Congestion Notification
BFWA	Broadband Fixed Wireless Access
BGP	Border Gateway Protocol
BSR	Bootstrap Router
CA	Certificate Authority
CBR	Constant Bit Rate
C-BSR	Candidate Bootstrap Router
CCC	Circuit Cross Connect
CE	Customer Edge LSR/Router
CGMP	Cisco Group Management Protocol
CIDR	Classless Inter-Domain Routing
CLI	Command Line Interface
CLNP	Connectionless Network Protocol
CoS	Class of Service
CPU	Central Processor Unit

C-RP	Candidate Rendezvous Point
CSMA-CD	Carrier Sense Multiple Access-Collision Detection
CSPF	Constrained Shortest Path First
CVS	Concurrent Version System
DCE	Data Communication Equipment
DCN	Data Control Network
DE	Discard Eligible
DNS	Domain Name System
DoS	Denial of Service
DPT	Dynamic Packet Transport
DR	Designated Router
DRR	Deficit Round Robin
DSCP	DiffServ Code Point
DTE	Data Terminating Equipment
DUAL	Diffusing Update Algorithm
DVMRP	Distance Vector Multicast Routing Protocol
DWDM	Dense Wavelength Division Multiplexing
EAP	Extensible Authentication Protocol
ECN	Explicit Congestion Notification
ECT	ECN Capable Transport
EIGRP	Enhanced Interior Gateway Routing Protocol
E-LSP	Diffserv integrated LSP using EXP bits to represent DSCP
ESP	Encapsulating Security Payload
EXP	MPLS Experimental Bits
FC	Feasibility Condition
FD	Feasibility Distance
FEC	Forwarding Equivalence Class
FECN	Forward Explicit Congestion Notification
FSC	Fibre Switched Capable
FT	Fault Tolerant TLV
FTP	File Transfer Protocol
GDA	Group Destination Address
GMPLS	Generalized Multiprotocol Label Switching
GPRS	General Packet Radio Service
GPS	Generalized Packet Scheduler
GRE	Generic Router Encapsulation
ICMP	Internet Control Message Protocol
IEEE	Institute of Electrical and Electronic Engineers
IETF	Internet Engineering Task Force
IGMP	Internet Group Management Protocol
IGP	Interior Gateway Protocol
IGRP	Interior Gateway Routing Protocol
IIH	IS-IS Hello PDU
IP	Internet Protocol

IPS	Intelligent Protection Switching
IRR	Internet Routing Registry
IS	Intermediate System
IS-IS	Intermediate System to Intermediate System
ISO	International Organization for Standardization
IXP	Internet Exchange Point
L2TP	Layer 2 Tunneling Protocol
LDP	Label Distribution Protocol
LIR	Local Internet Registry
L-LSP	Diffserv integrated LSP using labels to represent different DSCP
LOL	Loss Of Light
LOS	Loss Of Signal
LSA	Link State Advertisement
LSC	Lambda Switched Capable
LSDB	Link State Database
LSP	Label Switched Path (MPLS)
LSP	Link State PDU (IS-IS)
LSR	Label Switching Router
MAC	Media Access Control
MBGP	Multiprotocol Border Gateway Protocol
MD5	Message Digest 5
MED	Multi-Exit Discriminator
MIB	Management Information Base
MPLS	Multiprotocol Label Switching
MRU	Maximum Receive Unit
MSDP	Multicast Source Discovery Protocol
MSP	Multiplex Section Protection
MTU	Maximum Transmission Unit
NAS	Network Access Server
NAT	Network Address Translation
NLRI	Network Layer Reachability Information
NSSA	Not-So-Stubby Area
NTP	Network Time Protocol
OCn	Optical Carrier 'times n'
OID	Object Identifier
OOB	Out Of Band
OSI	Open System Interconnection
OSPF	Open Shortest Path First
P	Provider LSR (core LSR)
PCR	Peak Cell Rate
PDU	Protocol Data Unit
PE	Provider Edge LSR (aggregation LSR)
PHB	Per Hop Behaviour
PIM-DM	Protocol Independent Multicast Dense Mode

PIM-SM	Protocol Independent Multicast Sparse Mode
PKI	Public Key Infrastructure
PMTU-D	Path MTU Discovery
PoP	Point of Presence
POS	Packet over SONET/SDH
PSC	Packet Switched Capable
PSTN	Public Switched Telephone Network
QoS	Quality of Service
RADIUS	Remote Authentication Dial-In User Service
RED	Random Early Detect or Random Early Discard
RIB	Routing Information Base
RIP	Routing Information Protocol
RIPE	Réseaux IP Européens (The RIR for Europe, the Middle East and parts of Africa)
RIR	Regional Internet Registry
RMON	Remote Monitoring MIB
RP	Rendezvous Point
RPF	Reverse Path Forwarding
RPR	Redundant Packet Ring
RPT	Rendezvous Path Tree
RR	Route Reflector
RSN	Robust Security Network
RSVP	resource ReSerVation Protocol
RSVP-TE	RSVP with TE extension
RTP	Reliable Transport Protocol
SA	Security Association
SAR	Segmentation And Reassembly
SCR	Sustained Cell Rate
SDH	Synchronous Digital Hierarchy
SLA	Service Level Agreement
SNMP	Simple Network Management Protocol
SONET	Synchronous Optical Network
SPF	Shortest Path First
SPI	Security Parameter Identifier
SPT	Source Path Tree
SRP	Spatial Reuse Protocol
SSH	Secure Shell
SSL	Secure Sockets Layer
SSM	Source Specific Multicast
STM-n	Synchronous Transport Module 'times n'
TACACS	Terminal Access Controller Access Control System
TCP	Transmission Control Protocol
TD	Topology Discovery
TDP	Tag Distribution Protocol

TE	Traffic Engineering
TED	Traffic Engineering Database
telco	Telephone Company
TFTP	Trivial File Transfer Protocol
TKIP	Temporal Key Integrity Protocol
TLS	Transport Layer Security
TLV	Type/Length/Value
ToS	Type of Service
TSC	Time Division Multiplexing (TDM) Switched Capable
TTL	Time To Live
UBR	Unspecified Bit Rate
UMTS	Universal Mobile Telecommunications System
USA	Unicast Source Address
VBR-NRT	Non-Real Time Variable Bit Rate
VBR-RT	Real Time Variable Bit Rate
VPLS	Virtual Private LAN Service
VPN	Virtual Private Network
VPWS	Virtual Private Wire Service
VRF	Virtual Routing and Forwarding Table
VSA	Vendor Specific Attribute
WDM	Wavelength Division Multiplexing
WDRR	Weighted Deficit Round Robin
WEP	Wired Equivalence Privacy
WFQ	Weighted Fair Queueing
WLAN	Wireless Local Area Network
WPA	WiFi Protected Access
WRR	Weighted Round Robin

Introduction

Today's Service Provider (SP) marketplace is a highly competitive environment in which it is important to make the most of the assets you have. The halcyon days of the late 1990s, when providers had vast amounts of money to spend on equipment and transmission services and spent it at a prodigious rate, are long gone. It is essential that network architects today can design IP and MPLS networks, which are flexible enough to expand and, possibly more importantly, can cope with new and profitable services being added to the network.

WHAT IS THIS BOOK ALL ABOUT?

As a network engineer working and learning my job in the mid-1990s I often wished I had a book that would help me understand the decisions being made by our design and development team. I wanted to join that team and, if I was to design scalable networks for my employer, I needed to have that understanding. Unfortunately, no such book existed. There were a number of books describing the operation and behaviour of various IP routing protocols (useful resources in themselves) but none which explained the principles of how to build and operate large networks. I was extremely fortunate to be able to learn from other more senior members of the various engineering groups and this book is the amalgamation of what I was taught and what I have learned through self-study and painful experience.

In this book I will examine the architectural and design principles that can be applied to designing and building scalable IP and MPLS networks. Each chapter in this book presents a particular aspect of the overall process of designing a scalable IP network and provides examples of configurations using both Juniper Networks JUNOS software and Cisco Systems IOS software. I have chosen these two vendors as theirs are by far the most widely used devices in the Internet today.

This book will not provide in-depth descriptions of the protocols associated with IP routing and switching. For those seeking an in-depth treatment of those protocols, there are a number of other books that provide an excellent reference (e.g. *Routing TCP/IP*, Volume 1 by Jeff Doyle, *Routing TCP/IP*, Volume 2 by Jeff Doyle and Jennifer Carroll, and *Complete Reference: Juniper Networks Routers* by Jeff Doyle and Matt Kolon).[1] There will be detailed discussion, where necessary, to explain the way particular features and protocol behaviours impact the scalability of a network.

WHO IS THIS BOOK FOR?

This book is intended to provide a guide to those designing IP networks. It will provide both a guide and a reference to network architects and engineers. In addition to those designing networks, it will also be useful to network operations staff who want to understand some of the principles being applied in the architecture and design of their network.

It is also hoped that this book will provide a valuable resource for people studying for the CCIE and JNCIE exams from Cisco Systems and Juniper Networks, respectively. While design does not actually form part of those exams, a solid understanding of the design process can only be beneficial to candidates. Also, while not referring directly to any other vendors' software or hardware, the principles described herein should be applicable to designing or building a network using any vendor's equipment.

While this book focuses on processes and mechanisms with respect to the design of service provider networks, much of the material in this book can be applied to the design of any IP network, irrespective of whether it is a service provider's network or an enterprise's network.

It is assumed that the reader is familiar with the basic principles of IP networking and the protocols associated with IP routing and switching.

SCALING: GETTING BIGGER AND DOING MORE

Scaling means different things to different people. Most will agree that it means getting bigger. Getting bigger is certainly an extremely important issue but it is important that in order to grow, it is not necessary to entirely redesign the architecture of your network. The initial design of your network will dictate how easily you can make subsequent changes.

---

[1] Jeff Doyle *Routing TCP/IP*, Volume 1, Cisco Press, ISBN 1578500418. Jeff Doyle and Jennifer DeHaven Carroll, *Routing TCP/IP*, Volume 2, Cisco Press, ISBN 1578700892. Jeff Doyle and Matt Kolon (eds) *Complete Reference: Juniper Networks Routers*, Osborne McGraw-Hill, ISBN 0072194812.

In addition, it is important to recognize that scaling does not just mean building bigger networks but also means being able to offer a wider range of services to your customers using your existing assets. In today's marketplace, it is no longer sufficient to provide a network that provides simple IP access. Customers are demanding new and different services (e.g. VPNs, CoS/QoS-backed SLAs, etc.). The ability to add new services to your network will be pivotal to your potential to grow your business and maintain (or move into) profitability.

DESIGNING A NETWORK FROM SCRATCH

Few of us are lucky enough to have the luxury of designing a network from scratch. For those of us who are, there are opportunities that are not available when trying to enlarge or extend the functionality of an existing network. In a new network it is possible, within financial constraints, 'to do the right thing'. For example, it is possible to create efficient IP addressing and efficient aggregation schemes. This can significantly reduce the size of the routing tables in all your routers. In hierarchical routing models in which the core routers see all routes, this is particularly important. Reducing the size of your routing tables has a positive impact not only on the memory required to maintain them but also on the processing power required to calculate routing updates.

SCALING AN EXISTING NETWORK

In the previous section, it was pointed out how much easier it is to start from scratch. However, that does not mean that it is not possible to achieve great improvements in scalability in an existing network. When scaling an existing network, you will have to deal with decisions you may have made earlier and, often more significantly, decisions made by your colleagues and predecessors. It is sometimes possible to reverse decisions but, more often than not, you will have to work within the constraints they impose.

FUNCTIONAL ELEMENTS OF SP NETWORKS

As mentioned earlier, this book is intended to be of value to all network engineers, irrespective of whether they work for a service provider (SP) or a company. However, the elements of the larger SP networks can provide a useful basis with which to demonstrate the design principles. Some, although not all, SPs have networks that are excellent examples of scalability. SP networks have, over the past decade or so, scaled at a phenomenal rate. Every possible mistake has been made at some time by one or more

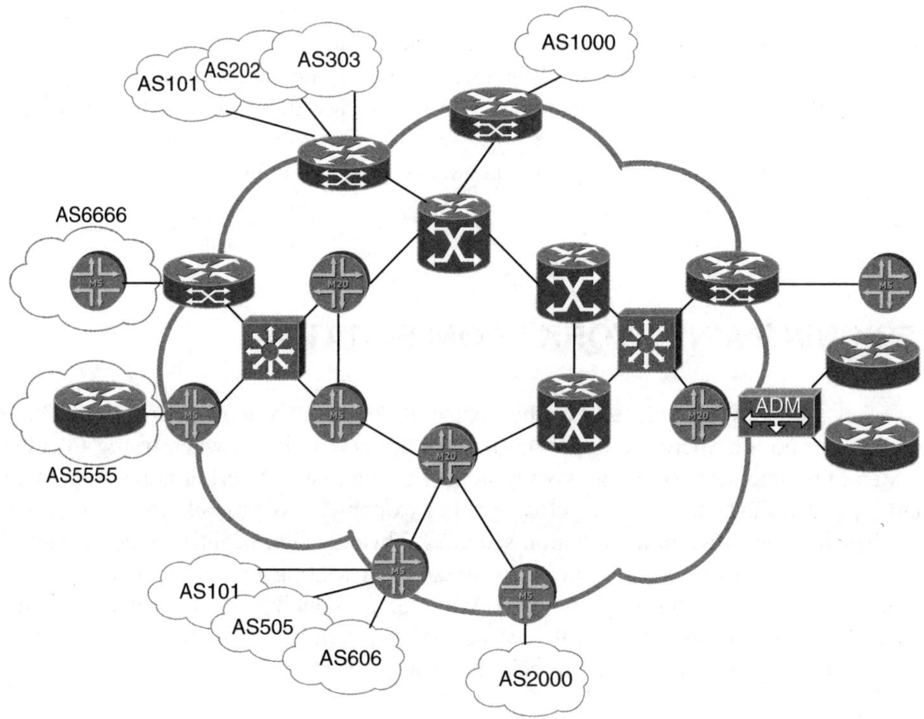

Figure I.1 Example network used in this book

SPs and every possible workaround to almost every possible problem has been tried on someone's network.

Throughout this book, we will be providing a number of examples. These will demonstrate aspects of scalability across the various functional elements of service provider networks. In order to do this, we will be using a single consistent network or autonomous system. This is shown in Figure I.1.

TERMS USED IN THIS BOOK

Various terms are used throughout this book to describe the different functional elements of SP networks. The terms used throughout this book are described below.

- *Core (Provider/P)*. A core switch/router is a device that connects only to other devices in the same network. It provides high speed, large capacity connectivity between access devices. Traditionally, core devices are relatively 'dumb' and do not provide much functionality beyond forwarding packets really fast. Modern core devices are able to support more features without any detrimental effects on performance.

Some SPs also use a further layer of distribution devices, perhaps to aggregate traffic from a regional network into the core or to aggregate traffic from particular types of access devices (e.g. CMTS) into the core.

- *Access (Provider Edge/PE)*. An access or edge switch/router is a device that provides connectivity between customer edge (CE) router on customer networks and the SP core network.

  - *Fixed access*. Fixed access refers to all access provided over fixed transport media. This includes leased lines (E1/T1, E3/DS-3, STM-x/OCy), metro Ethernet, fiber to the home/curb, etc.

  - *Broadband (Cable/DSL)*. Technically, this is a subsection of the previous section since almost all broadband services are currently provided over fixed lines (cable or telephone lines). There are, however, some wireless DSL services. However, there are particular aspects of these access methods, which require specific consideration.

  - *Wireless*. This covers all forms of non-fixed access (GPRS/UMTS, WLAN, BFWA, etc). Wireless access presents a number of specific problems associated with security, capacity and reliability.

- *Border (Provider Edge/PE)*. This is a switch/router that is somewhat like an access device. However, it provides access to other provider networks rather than to customer networks. This is a small but incredibly important difference and has a major impact on the configuration requirements.

  - *Peer networks*. Peer switch/routers are border devices that connect your network to peer networks. Peer networks are other provider networks with which you want to exchange routing information while sharing the cost of the connection. No money normally changes hands between the operators of networks with a peering relationship.

  - *Transit*. Transit switch/routers are border devices that connect your network to your upstream transit provider. Transit providers are other provider networks, which provide connectivity to the Internet or, in some cases, a particular region (e.g. Europe, America, Asia). This is a customer/provider relationship and the transit provider will charge a fee for this service.

- *Customer Edge (CE)*. This is a switch/router that provides connectivity between a customer network and the PE on a SP network. In some respects, the Transit PE described above could be viewed as a CE device. As far as the transit provider PE is concerned, that router is a CE.

ORGANIZATION OF THIS BOOK

This book is divided into a number of chapters, each of which deals with a particular aspect of designing an IP/MPLS network.

CHAPTER 1 HARDWARE DESIGN

In this chapter, I will take a look at some of the different choices made by hardware vendors in the design of their equipment and how each of their choices might impact the scalability of your network. It might seem at first that this is entirely out of your hands. However, an understanding of the architecture of vendors' equipment can be exceptionally useful in enabling you to scale your IP network.

CHAPTER 2 TRANSPORT MEDIA

This chapter looks at the various transport media available and evaluates the impact on scalability of each of them. Different media have significantly differing levels of overhead just in the transportation of the packets. They also impose different loads on the network in respect of maintaining the link. It is also important to note that the routing protocols you choose and their behaviours may influence your chosen transport media, and vice versa. Routing protocols are discussed in Chapter 5.

CHAPTER 3 ROUTER AND NETWORK MANAGEMENT

This is an area which is often overlooked but can dramatically affect the administrative effort required to operate your network as it grows. Having the tools to understand and modify the behaviour of your network is essential. The greater your understanding, the more accurate your ability to modify your network behaviour.

CHAPTER 4 NETWORK SECURITY

Security is clearly a significant issue for all networks, large or small. As your network grows, the security issues become bigger. It is also an unfortunate fact of life that the frequency, variety and impact of attacks have grown significantly over the past few years. The details of these attacks are often made public and implementations published on the Internet. This makes it easy for anyone with the inclination to (effectively anonymously) mount an attack on your network from hundreds or even thousands of globally distributed sources. The proliferation of 'always on' connections to home users has provided these attackers with millions of potential targets and sources.

If you have a consistent and carefully prepared security plan, security will not hinder the scalability of your network. Without such a plan, you will likely spend a lot of time dealing with attempts to breach the security of your network and that of your customers. This time costs your company money and could be more profitably spent implementing other services.

CHAPTER 5 ROUTING PROTOCOLS

This chapter covers all the main unicast routing protocols used in SP networks today. The scalability of your network is fundamentally reliant upon the scalability of your chosen routing protocols and, equally importantly, on the way in which you configure them. No matter how excellent the inherent scalability of the routing protocols you use, if you configure them poorly, your network will not scale, in any sense of the word.

CHAPTER 6 ROUTING POLICY

Routing policy is the technical means by which you implement your business policies with respect to your customers, peers and transit providers. As with the routing protocols themselves, it is critical that you develop effective routing policy to assist you in scaling your network.

CHAPTER 7 MULTIPROTOCOL LABEL SWITCHING (MPLS)

Multiprotocol label switching was initially devised as a means of improving the speed of IP forwarding (actually by doing a lookup on a fixed length label rather than a variable length IP prefix) in software-based routers. As vendors built hardware (ASIC) based routers, it became clear that the lookup performance of a variable length IP prefix was equal to that of a fixed length MPLS label. No longer needed for improvements in forwarding performance, the traffic engineering capabilities of MPLS became the focus. Traffic engineering allows network operators to make more effective and efficient use of their transmission assets.

CHAPTER 8 VIRTUAL PRIVATE NETWORKS (VPNs)

One of the major 'fashionable' services being offered to customers is IP-based Virtual Private Networks. It is suggested that VPNs can save vast amounts of money for enterprises when compared to the cost of private lines or even an ATM or Frame Relay-based VPN. This may indeed be the case for the enterprise but if the service is not scalable, then it is almost certain to be unprofitable and unmanageable for the service provider.

CHAPTER 9 CLASS OF SERVICE AND QUALITY OF SERVICE

The issue of CoS/QoS is sometimes closely associated with VPNs but can be applied to all network traffic. This association seems inevitable given that IP VPNs are often presented as an alternative to ATM VPNs with all of their inherent QoS functionality.

Adding CoS/QoS to an IP/MPLS network is a significant undertaking. It is important to understand what customers want and it is important to strike a balance between the perfect service as perceived by customers and a realistic, scalable service from the point of view of the service provider.

CoS/QoS provides the mechanism whereby SPs can offer differentiated services for which they can charge a premium. As the number of services running over a single infrastructure increases, so the importance of CoS/QoS increases. It is essential that best-effort traffic generated by low-paying customers must not overwhelm priority traffic generated by premium customers.

CHAPTER 10 MULTICAST

Multicast design can be a difficult thing to get right. One of the goals of multicast is to make the transmission of identical information to multiple receivers much more scalable. However, if used in the wrong circumstances or poorly implemented in your network, it can be detrimental to the overall scalability of all aspects of your network operations.

CHAPTER 11 IPv6

IPv6 is the latest version of the Internet Protocol. It has a number of enhancements over IPv4, which include improved security, improved mobility and vastly extended address space. This chapter will cover the differences between IPv4 and IPv6 including the modifications to the IP frame format, routing protocols and control protocols. It will also look at how the requirements imposed by IPv6 in the future can impact your network design decisions today.

CHAPTER 12 COMPLETE EXAMPLE CONFIGURATION FILES (IOS AND JUNOS SOFTWARE)

This chapter will provide complete configuration examples using both Cisco Systems IOS and Juniper Networks JUNOS software. These configurations help to combine the elements of this book and present them in a real-world format that displays each of the major elements of an SP network.

1

Hardware Design

Hardware design may, at first sight, seem to be something over which the network architect or designer has no control. It would appear, therefore, that it plays very little part in the design of a scalable Internet Provider (IP) network. Despite this, an understanding of the various architectures employed by the hardware vendors can influence the choice of hardware. Thus, the choice of hardware will inevitably affect your network designs.

There are only a finite number of architectures on which hardware designs are based. Most vendors of core Internet routers today use various Application Specific Integrated Circuits (ASIC) and/or network processors designed specifically for the purpose of performing IP and MPLS lookups, applying access control/firewall functionality, enforcing policy and applying Quality of Service (QoS). However, there are significant differences in the various implementations of architectures within the plethora of available access routers and switches. Some still rely on a software-based system with a more general-purpose processor, which provides the flexibility to add services at the expense of forwarding performance. This contrasts with ASICs, which are (as the name suggests) designed for a single function, or group of functions. The first generation of ASIC-based devices provided very high-speed forwarding at the cost of flexibility and services. In addition, this improved performance is gained at the cost of higher development cost and longer time to market. The introduction of network processor-based systems has combined the flexibility of a general-purpose processor without sacrificing any of the packet forwarding capabilities of an ASIC. In addition, the non-specific nature of the network processor can lead to lower costs than those associated with ASICs.

Designing and Developing Scalable IP Networks   G. Davies
© 2004 Guy Davies   ISBN: 0-470-86739-6

1.1   SEPARATION OF ROUTING AND FORWARDING FUNCTIONALITY

One of the major steps forward in ensuring scalability in network performance over the past few years has been the move towards the separation of routing and forwarding functionality. Put simply, the process of calculating the forwarding table is run on different hardware from that responsible for actually implementing the packet forwarding.

By separating these functions onto different hardware modules it is possible to continue forwarding packets at line rate even in the presence of severe disruptions to the network topology. Designs based on general-purpose processors have to use processor cycles to make both the routing decisions and the forwarding decisions. Therefore, if the network topology is dramatically and repeatedly updated, the processor with a fixed number of cycles is required to take cycles from the forwarding tasks to carry out the routing tasks or vice versa.

1.2   BUILDING BLOCKS

A router based on the principles of separation of routing and forwarding functionality is based on a few simple building blocks, (see Figure 1.1).

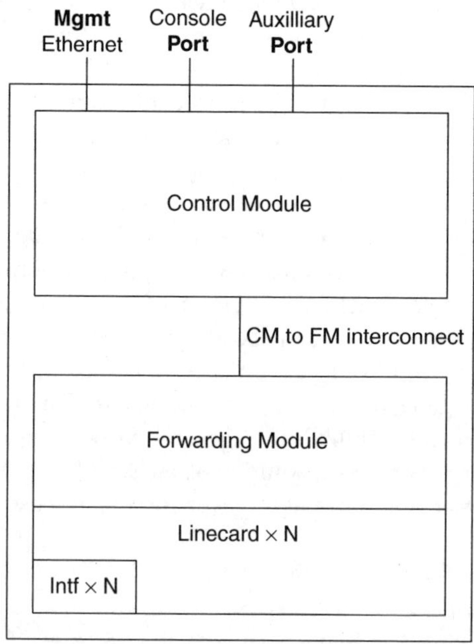

Figure 1.1   The logical architecture of an IP router (separate forwarding and routing planes)

1.2.1   CONTROL MODULE

The control module is responsible for running the routing protocols, creating a routing table from the sum of the routing information received and generating a forwarding table from the routing table. This includes the enforcement of policy relative to the routing information being imported and exported. In addition to the routing function, this module provides Command Line Interface (CLI), management and logging functionality.

1.2.2   FORWARDING MODULE

The forwarding module is responsible for the receipt of packets, decoding the Layer 1, Layer 2 and Layer 3 information and sending that packet to the appropriate output interface (or interfaces in the case of multicast) where it is re-encoded with Layer 3, Layer 2 and Layer 1 information before transmission onto the wire. The forwarding module is also the ideal location for implementing access control/firewall filters and class/quality of service. ASIC and Network Processor-based devices tend to implement these features in the forwarding module. Traditional, software-based routers always had to send every packet to the CPU if any of these features were enabled. This was one of the main reasons why software-based routers were unable to implement complex services while maintaining high packet throughput.

1.2.3   NON-STOP FORWARDING

Non-stop forwarding is the generic term for the ability of a router to carry on forwarding packets even though the device responsible for the calculation of the routing and forwarding tables is temporarily inoperable. The specific reason for the inoperability of the routing device is really of no importance at this stage of the discussion. Non-stop forwarding fundamentally relies on the separation of routing and forwarding functionality within the router. In addition, each routing protocol requires a number of modifications to the protocol operation in order to prevent the clearance of routes associated with the failed protocol from the forwarding table on the forwarding modules, both on the local router and on the neighbours of the local router, while the routing process or module recovers.

1.2.4   STATEFUL FAILOVER

This takes non-stop forwarding concept one step further. In a switch or router with redundant control modules it is possible to duplicate the state associated with routing protocols from the master routing module to the slave routing module. The slave control module is held in a state in which it can 'instantly' take over the operation of a failed master control module. Generally, the slave can take over operations within a matter of seconds.

There are, as usual, strongly held opinions about whether this is a good idea or not. Some vendors feel that it is foolhardy to copy all the state from the active control module to the slave. The main concern is that if a particular software bug or combination of state causes the master control module to reset and that state has been copied faithfully onto the slave, then, in all likelihood, the same problem will arise on the slave and cause it to reset. This would clearly be something of a disaster, given that the whole point of mirroring the state onto the slave control module is to prevent the total failure of the device.

On the other side of the argument, this mechanism provides the potential for the fastest failover. By maintaining an exact copy of the state from the master control module on the slave control module, it is possible to begin forwarding packets immediately without having to rebuild the routing information bases (RIB). This mechanism imposes no extra requirements upon the routing protocols to enable non-stop forwarding.

1.3 TO FLOW OR NOT TO FLOW?

This might sound like a rather bizarre question. The obvious answer is 'of course we want the traffic to flow'. But this question does not actually relate to the flow of traffic but to the internal architecture of the network device.

Many Layer 2 switches work on flow-based mechanisms. A flow is a single stream of traffic associated with a single 'conversation' between two network devices. These can be uniquely identified by the source and destination address and source and destination TCP or UDP port. When the first packet for a new flow arrives, the packet is sent to the CPU and the switch creates a new entry in a flow table, which identifies the flow characteristics along with the ingress and egress ports. Once the flow entry has been created, then all subsequent traffic matching that flow is switched from the ingress to the egress port without touching the CPU, thus dramatically improving the performance. In an environment in which there are a constrained number of long-lived flows, these devices provide extremely good performance. However, if there are a large number of short-lived flows, performance becomes constrained by the ability of the CPU to create new entries in the flow table and by the amount of memory available to store the flow table. If there are too many new flows, either new ones will not be created or existing ones will have to be purged from the flow table to make way for the new ones. As soon as another packet for the purged flow arrives, it requires the set-up of a new flow. Thus, the problem is exacerbated.

As the line between Layer 2 devices and Layer 3 devices became blurred, many flow-based switches acquired Layer 3 (and Layer 4 through 7) functionality. In an enterprise environment, this is not a significant issue because flows are generally fewer in number and longer-lived. However, in a service provider there are almost invariably vast numbers of comparatively short-lived flows, particularly at major aggregated egress points. This is the worst possible situation for flow-based devices.

I have experienced instances of service providers installing flow-based devices at an Internet Exchange Point (IXP) and, very quickly, replacing those routers with routers of traditional design because of the intolerable performance of those devices in that role. You may be asking what the attraction of such devices was given their clear disadvantages within a service provider network. The simple answer often is cost. Switch-based routers (a.k.a. Layer 3 switches) are invariably cheaper than their traditional counterparts, sometimes by a factor of three or four. They also generally have hybrid functionality so they can be used in both Layer 2 and Layer 3 roles. These advantages in terms of price and flexibility are often sufficient to convince hard-pressed service providers to try to get more for their money.

To be fair to the vendors of flow-based routers, many have enhanced their devices with much more powerful CPUs and much more memory for the flow tables. This helps to mitigate some of the problems but does not solve the inherent issues. The problem is just masked until the number of flows or rate of creation of flows grows sufficiently to overwhelm the new hardware.

One mechanism for overcoming this issue is tunnelling, since the packets are forwarded based on the outer headers, which relate only to the end points at the edge of the service provider's network. One widely used implementation of this feature is MPLS (discussed more fully later in the book). Because tunnels or Label Switched Paths (LSPs) are built between the edge routers in the networks and all traffic between those points is carried within the LSP, it aggregates to a single 'flow'.

Even with the enhanced devices, you should think long and hard before you use a flow-based router for a service provider network. Think particularly about the environment in which the router will be used and be sure that the performance constraints of the flow-based router will not come back and bite you.

1.4   HARDWARE REDUNDANCY, SINGLE CHASSIS OR MULTI CHASSIS

The choice of mechanism used to provide redundancy for each function within a single network node (a.k.a. hub) usually arouses one of two generally strongly held opinions. One theory is that it is best to create a single network element in which all the major service-impacting elements are replicated. This works on the premise that you generally provide $N + 1$ ($N >= 1$) processing, switching, forwarding/interface, power and cooling modules so that if one module fails, the router can continue to function without any degradation in service. As stated above, failure of the control modules can have a brief impact on the normal routing behaviour of the device but forwarding should operate undisturbed (assuming certain preconditions are met). Failure of the switching fabric, the internal 'infrastructure' over which packets pass when travelling between linecards, should not have any impact on forwarding with the exception of the loss of any packets actually traversing the failed fabric at the instant it fails. Failure of forwarding/interface modules is more difficult to do without any impact because there usually has to be some element of external support for this (i.e. the far end has to also support the same feature—APS, dual PHY, etc.). This approach also means that there is generally a requirement for less power and rack space. There is also a reduction in the demands placed on your routing protocols. Fewer links and fewer nodes mean less state and fewer changes in topology. Finally, the reduced administrative and support costs of operating a single device rather than several devices can be significant.

An alternative theory is that no matter how clever the design (or how much money you spend), it is not possible to produce a single unit with sufficient resilience in all the aspects (particularly forwarding) to adequately protect customer traffic. Therefore, it is better to build a resilient function by provisioning devices in groups, so that the complete failure of a single device may lead to reduced performance but never total loss of connectivity. However, more chassis inevitably mean a greater demand for power, more real estate, more instability in the network topology and greater support costs.

As with most things, designing a network is full of compromises and what usually happens is that service providers end up implementing a combination of the two theories. Traditionally, core routers, in which the forwarding function is paramount, are often provisioned in pairs with each taking a share of the links from other core sites and each having an equal capacity into the hub. The loss of one core router, therefore, reduces the aggregate capacity into the hub but does not isolate the hub from the rest of the network. The negative side of this model is that it significantly increases complexity. Due to the need to interconnect these boxes and to connect all edge boxes to both core routers, it inherently increases the number of interfaces required to service a single hub, thus increasing capital costs. Also, the added maintenance associated with more interfaces means greater operational costs. As the support for non-stop forwarding improves, it becomes more realistic to have single core routers in the core, thus dramatically reducing costs. As always, this is a big positive. The main issue is one of trust. If there is only one core router connecting a PoP to the rest of the network, the risk associated with the loss of that device is related to the number of edge routers, and therefore customers, connecting to the core via that single router. The question you have to ask yourself is whether you trust a router enough to be able to rely upon it to meet your SLAs. For instance, if one software upgrade requires a 6-minute reload, that would invalidate a 99.999% uptime SLA.

With edge routers, particularly access routers to which customers often have a single connection, it is not usually possible to usefully deploy them in pairs. Therefore, more of the single chassis redundancy features have been applied more widely to edge routers. If customers have only a single connection to a single router, it is essential that each router has resilient power, cooling, switching and control modules and, as far as is possible, resilient forwarding modules. The same risks to an SLA apply as above. However, the number of individual customers which a single failure can affect is generally significantly lower for edge routers because core routers typically aggregate several edge routers and their associated customers, thus, the impact is much greater with core routers.

2

Transport Media

The choice of transport media may, at first sight, appear to be a relatively arbitrary decision based more on issues of convenience and functionality than on scalability. However, the choice of transport media, and the architecture within which it is used, can have a significant impact on the scalability of the network.

This chapter will give a brief description of the features of each of the transport media and any scalability issues associated with carrying IP traffic. In some cases, these scalability issues are more generic, affecting Layer 3 protocols when using that Layer 2 technology. In other cases, the scalability issues are specific to the transport of IP.

There are a number of issues which are common to more than one transport media. We will take a look at these independently from the specific transport media.

2.1 MAXIMUM TRANSMISSION UNIT (MTU)

The MTU, as the name suggests, is the largest frame that can be transmitted across a particular link. In the original Internet, packets were transmitted with an MTU of 576 bytes or the MTU of the directly connected network, whichever was the lower. An MTU of 1500 bytes is now the default on Ethernet-based media with much higher MTUs possible on most media. If a router receives a frame on one interface, which has a frame size greater than the MTU of the outgoing interface, then it must split the packet into two or more packets so that each packet is small enough to be transmitted. This process of fragmentation is computationally expensive and may require that the packet being fragmented be sent to the central processor. Packets are not rebuilt until they arrive at the destination host.

Designing and Developing Scalable IP Networks   G. Davies
© 2004 Guy Davies   ISBN: 0-470-86739-6

Sending packets with artificially low MTUs is inefficient and can lead to significantly lower throughput than could be achieved if the optimum MTU is used. This is because the IP and MAC headers are effectively a fixed size and as the MTU drops, the IP and MAC headers form a larger proportion of the total data transmitted over the wire.

This leads to the conclusion that fragmentation is a bad thing. As such, the ideal situation is to avoid fragmentation altogether. One means of assuring that fragmentation is not used is simply to ensure that the core of the network uses an MTU higher than that used at the edge. Almost all end hosts are attached to media with MTUs less than or equal to 1500 bytes. However, it is common to use technologies in the core of the network, which support MTUs in excess of 4400 bytes and some support MTUs of more than 9000 bytes.

2.1.1 PATH MTU DISCOVERY

A more elegant technical solution to this problem is the use of Path MTU Discovery. This mechanism works by transmitting packets with the MTU set to the MTU of the directly connected link and the Don't Fragment (DF) bit set. If any link along the path has an MTU smaller than that used to transmit the packet, an ICMP Destination Unreachable message would be returned to the sender. This message includes a field, the Next-Hop MTU, that includes an indication of the MTU of the hop across which the entire packet could not be transmitted without fragmentation. Upon receipt of the ICMP Destination Unreachable packet, the source must reduce the MTU in line with the Next-Hop MTU. This process ends when either the packets start reaching their destination or if the source host decides to stop further reducing the MTU and clears the DF bit in subsequently transmitted packets.

This process is totally reliant upon the generation of ICMP Destination Unreachable messages. However, ICMP Destination Unreachables can be used as a means of generating a denial of service attack on routing devices because the generation of ICMP messages takes up processor cycles, which would otherwise be used for other functions. Because of the threat of denial of service attacks, some network operators configure their routers not to send ICMP Unreachables. This breaks Path MTU Discovery.

Path MTU Discovery is particularly important for IP security (IPsec), MPLS and other tunnel-based services. In many cases tunnels do not tolerate the fragmentation of packets so it is essential to ensure that excessively large packets are not injected into the tunnel.

2.1.2 PORT DENSITY

This could be viewed as part of the hardware chapter. However, since different transport media are presented via different copper and fibre presentations, the issues can be usefully discussed here.

There are a variety of connectors associated with both copper and fibre presentations. Copper is presented via RJ45, RJ48, various BNC type connectors, X.21 among others. Fibre has a similarly wide array of connectors including SC, ST, FC-PC, LC and MT-RJ.

This array of different connectors has, to some extent, grown out of the need to get more and more ports securely connected to the minimum real estate on the equipment. The choice of connector is often not made by the network operator since they are constrained by the choices made by the hardware vendors. However, you may find that there is a choice between fibre or copper. This is a choice which is often made on the basis of convenience of connecting different vendors' devices.

2.1.3   CHANNELIZED INTERFACES

A neat solution to the demand for increasing port density is the use of channelized interfaces. These interfaces use the fact that traditional telephone companies' (telcos) transmission networks often carry lower bandwidth circuits in larger capacity aggregates. The transmission networks allow the adding and dropping of the lower capacity circuits onto the aggregates at arbitrary points on the network. It is, therefore, possible to take tributary circuits from a wide geographical area and aggregate them into a single aggregate. Traditionally, those tributary circuits would be broken back out before being presented to the ISP. However, with the availability of channelized ports for routers, telcos can now deliver the aggregate direct to the ISP and the ISP can support many circuits on a single physical port (e.g. 61 E1s on a single STM-1, 12 DS-3s on a single OC-12). Now, with a single line card on a router capable of carrying several channelized STM-1s, for example, it is possible for thousands of customers to be supported on a single device or even a single linecard. This raises serious questions about the reliability of a network device, when thousands of customers can lose service if it fails (see Chapter 1 for a more thorough discussion of this issue).

2.2   ETHERNET

Ethernet, in all its forms, is considered to be a comparatively cheap transport media. Traditionally considered a Local Area Network media, it is now being extended into the Metro Area and even the Wide Area network through use of technologies such as WDM transport over fibre using Long Haul optics.

Ethernet has developed from a bus-based architecture into a collapsed bus based on hubs and subsequently into a star-based architecture based on switches. The bus architectures were constrained to distances of 185 metres for 10Base2 and 500 metres for 10Base5. These are (almost) never used in service provider networks today. With the introduction of hubs and switches, the distance between the node and the hub on each link was limited to 100 metres. Only with the advent of Fast Ethernet and Gigabit Ethernet over fibre did the distance over which Ethernet could be used increase to the current situation, where it is possible to use long-haul optics to drive point-to-point dark fibre up to around 70 km without amplification. Cheaper, short-haul optics can more commonly transmit up to 250 to 500 metres over multimode fibre, depending upon the properties of the fibre.

Fundamentally, Ethernet is a broadcast medium. This is exemplified by the original bus architecture of Ethernet. It uses Carrier Sense Multiple Access with Collision Detection (CSMA-CD) as a transmission brokerage mechanism. Any node wishing to transmit must first listen to the media to see if any other device is currently transmitting. If no transmission is heard, then the node goes ahead and transmits its traffic. If transmission is heard, the node waits a randomized time before listening again. If two nodes start transmitting simultaneously, then both will detect the collision and transmit a 'scrambling' message to ensure that every receiver is aware of the collision. They then wait for a randomized time before trying to transmit again.

On bus-based systems (including the collapsed bus in a hub), there is a genuine collision domain, which incorporates all attached devices. Since there is no way that two devices can simultaneously transmit and we can assume that no-one is interested in listening to their own transmissions, these bus-based networks are inherently half-duplex in nature. Switches reduce the collision domain to the single link between the switch and each attached node. In addition, each attached device is connected with separate transmit and receive paths, meaning that incoming packets can never collide with outgoing packets. Since the device at each end of the link is the only device transmitting on a particular wire, they can transmit without first checking and be certain that there will never be a collision.

In addition to the ability to transmit full duplex across a single link, switches provide the means to transmit many simultaneous parallel streams of data. This means that it is possible to obtain true throughput of several gigabits per second, often only constrained by the capacity of the link between line cards and the backplane of the switch.

The ability to transmit full duplex on each link provides the basis for full duplex transmission end to end. As the number of hosts in a single collision domain grows, the number of collisions will inevitably increase. As the number of collisions increases, the throughput of valid data decreases dramatically. Switches, therefore, have a vital part to play in the scalability of Ethernet-based networks.

All the benefits of full duplex communication associated with switches are also available when Ethernet is used in a point-to-point mode. The same distance constraints also apply, 100 metres with a crossover Unshielded Twisted Pair cable and between 500 metres and around 70 kilometres with fibre, depending on the optics driving the line.

2.2.1 ADDRESS RESOLUTION PROTOCOL (ARP)

In order to carry IP over Ethernet, it is necessary to map IP addresses to Ethernet MAC addresses. This is achieved by the use of ARP. ARP is a relatively easy protocol, which is based on two simple messages. The first message is an ARP Request, also known colloquially as 'who has'. This message is sent by hosts wishing to map a known IP address to an unknown MAC address. A 'who has' message containing the known IP address is broadcast onto the segment. The second message type is an ARP Reply, also known colloquially as 'I have'. This is unicast directly to the requester and contains the IP address, along with the MAC address of the responder. In addition to replying with its own details, the responder will also record the IP address and MAC address of the requester into its own mapping table. This mechanism raises yet another problem. Since ARP uses broadcast messages, they are received by all hosts on the segment, irrespective of whether the recipient has the requested IP address. As the number of hosts on the network grows, the number of ARP messages will inherently grow. Since ARP messages are broadcast, they are received by every node on the segment. As the number of nodes grows, the number of unnecessary packets arriving at each node will grow. In large Ethernet networks with a very flat hierarchy, this can easily grow to become a significant burden. It is generally held that single segments of more than one thousand nodes are vulnerable to disruptions due to the massive number of ARPs sent during the standard transmission process. If there are sufficient hosts, the number of ARPs can actually grow to the point where they actually consume a significant proportion of the available bandwidth in 'ARP storms'.

2.2.2 MTU

Ethernet was initially constrained to a maximum MTU of 1500 bytes but now there are implementations with maximum MTUs in excess of 9000 bytes, which are known as jumbo frames. Jumbo frames bring with them both benefits and disadvantages. The advantages of jumbo frames are fairly obvious. If the large MTUs are available right through the core, then your network devices do not have to perform any fragmentation. Within large networks that carry large numbers of route prefixes in the core routing protocols, this can significantly improve the efficiency of the routing protocol exchanges.

The main disadvantage associated with jumbo frames is fragmentation if the MTU along the path is not consistently large and path MTU discovery is, for whatever reason, not successful.

2.3 ASYNCHRONOUS TRANSFER MODE (ATM)

ATM is a versatile transport protocol, which has been used extensively by service providers as a flexible high-speed transport with excellent traffic engineering and QoS functionality. ATM is not like other protocols described here, because it is carried over another transport media, e.g. SONET, E3, etc. In this respect, ATM is more like PPP or HDLC (i.e. a Layer 2 protocol) rather than the other media, which operate at Layer 1. However, as with the other transport media, ATM requires IP packets to be encapsulated in a sub-protocol. In the case of ATM, this encapsulation layer is called the ATM Adaption Layer 5 (AAL5). This, along with the fixed cell size and associated padding of incompletely filled cells can make ATM exceptionally inefficient, e.g. one 64-byte IP packet is carried in two 48-byte cells, the second of which contains 32 bytes of padding. Then, in addition, you have 10 bytes of ATM cell header (2 cells) and also the 8 bytes of the AAL5 trailer. This gives a total of 50 bytes of overhead to carry a 64-byte packet (>43% overhead). The absolute worst case is for a 49-byte IP packet, which results in

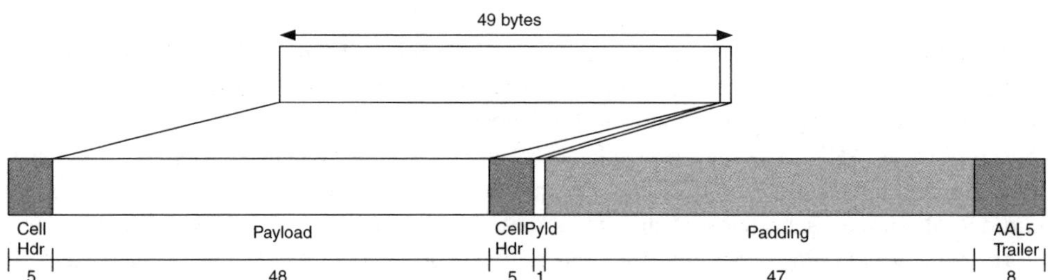

Figure 2.1   Overhead associated with the transport of IP packets in ATM cells

65 bytes of overhead $(47 + 5 + 5 + 8)$ for the carriage of 47 bytes of data, an incredible 57% overhead. Clearly, this is the worst case, see Figure 2.1. On average, ATM tends to suffer from around 20–25% overhead. When considering transcontinental multi-megabit circuits, a loss of 20 to 25% is extremely expensive.

ATM suffers from one other limitation, which constrains its long-term scalability. The process of splitting up the IP frames into 48-byte chunks for installation into the ATM cells and then taking those 48-byte chunks and rebuilding the complete IP packet is known as Segmentation And Reassembly (SAR). This is a computationally expensive function. The expense of building a module with sufficient processing and memory means it is financially unrealistic to create a SAR for widespread use that operates at greater than 622 Mbps. This means that it is not possible to have a single flow of data exceeding 622 Mbps. While that might seem like an extremely large flow of data, when considering macro flows between two major hubs on a large service provider's network, this is not excessive.

Given all these downsides, you might be wondering why anyone in their right mind would choose to use ATM as a transport for IP packets. There must have been some reasons why, in the mid to late 1990s, many of the largest service providers in the world relied upon ATM in their backbones. There were, of course, excellent reasons, not least of which was scalability! Prior to ATM, many large ISPs had used multiple DS-3s and Frame Relay switches to create an overlay network. However, as the flows of data grew, it became necessary to run more and more parallel links between each pair of hubs. At the time, DS-3 was the largest interface available on frame relay switches. ATM was the obvious next step for the service providers in need of greater capacity on individual links since ATM switches had interfaces running at OC3 and OC12 and could be used in a familiar overlay scheme. ATM also has great qualities for traffic engineering. This allowed ISPs to make better use of expensive bandwidth, and efficiently reroute traffic around failures and bottlenecks.

However, as networks inevitably continued growing, the SAR limitation became significant. With the largest available SAR being 622 Mbps, it was necessary to connect routers to a switch with several links in order to carry sufficient traffic. In the late 1990s, the largest service providers started building new core networks using Packet over SONET and MPLS. This combination provided many (but not all) of the benefits of ATM without the constraint of requiring SAR.

While ATM is certainly not a scalable solution in the core of the larger, global service providers, it remains a highly effective (although not particularly efficient) transport media for small to medium-sized ISPs and for medium to large enterprises with moderately large networks.

2.4   PACKET OVER SONET (POS)

In this classification, we include not only Packet over SONET/SDH but also Packet over wavelength or Packet over dark fibre, which also use SONET/SDH framing. POS has been widely used since the late 1990s for Wide Area circuits, particularly in service providers' backbones. POS is highly efficient in comparison to ATM as a transport for IP packets. Rather than ATM and AAL5 encapsulation, POS encapsulates IP within HDLC or PPP encapsulated within HDLC. An OC-3 circuit running POS can transport around 148 Mbps of IP data out of 155 Mbps compared to around 120 Mbps of IP data on an identical circuit running ATM. This vastly improved efficiency made POS extremely attractive to operators contemplating using STM-16/OC48 (2.5 Gbps) circuits and above. While ATM switches were capable of supporting STM-16 circuits between themselves, the constraints on SARs mean that it is still only possible to connect a router to the ATM switch at a maximum of STM-4/OC-12. In addition, the prospect of losing 20% of 2.5 Gbps as pure overhead was extremely unpalatable. Losing 20% of an STM-64/OC-192 was considered totally intolerable.

However, the flip side was that POS lacked any of the traffic engineering and QoS functionality available with ATM. It was only with the advent of MPLS that the lack of traffic engineering (and, more recently, the lack of QoS functionality) have been overcome to a certain degree. This removed one of the major objections of some of the engineers at the largest ISPs to using POS and paved the way for the use of STM-16 and STM-64 circuits.

2.5   SRP/RPR AND DPT

Spatial Reuse Protocol (SRP) was originally developed by Cisco in the late 1990s. It is a resilient, ring-based MAC protocol, which can use a variety of Layer 1 media but, almost invariably, is currently implemented using SONET/SDH encapsulation. This protocol has been documented in an informational RFC (RFC 2892) and adopted by the IEEE to produce Redundant Packet Ring (RPR) 802.17. Dynamic Packet Transport (DPT) is Cisco's implementation of SRP/RPR.

DPT/RPR is based upon a dual, counter-rotating ring. This provides the basis for the efficient (re)use of bandwidth between various points on the ring. Each node on the ring learns the topology of the ring by listening for Topology Discovery (TD) packets. The TD packets identify the ring on which they were transmitted. This, along with the list of

MAC addresses allows hosts to identify whether there has been a wrap of the ring (see below for a further explanation).

SRP can use mechanisms associated with the Layer 1 functionality (e.g. Loss of Signal (LOS), Loss of Light (LOL) with SONET/SDH) to identify the failure of links and nodes. However, since it is not constrained to media with this functionality built in, it is necessary to include a keepalive function. In the absence of any data to send, a router will transmit keepalives to its neighbour.

As can be seen from Figure 2.2, it is possible for several pairs of nodes to communicate at full line rate, simultaneously, without any interference. However, this relies upon the communicating nodes not having to pass through other nodes and no other node needing to communicate with yet another node. For example, R1 and R2 can communicate with each other and R3 and R4 can communicate with each other. However, if R1 and R4 wanted to communicate and R2 and R3 wanted to communicate, they would have to share the bandwidth between R2 and R3.

SRP uses some Cisco-patented algorithms to ensure fairness of access to the ring. These prevent traffic on the ring from starving a particular node of capacity to insert traffic onto the ring or vice versa. A full description of the algorithms is included in RFC 2892.

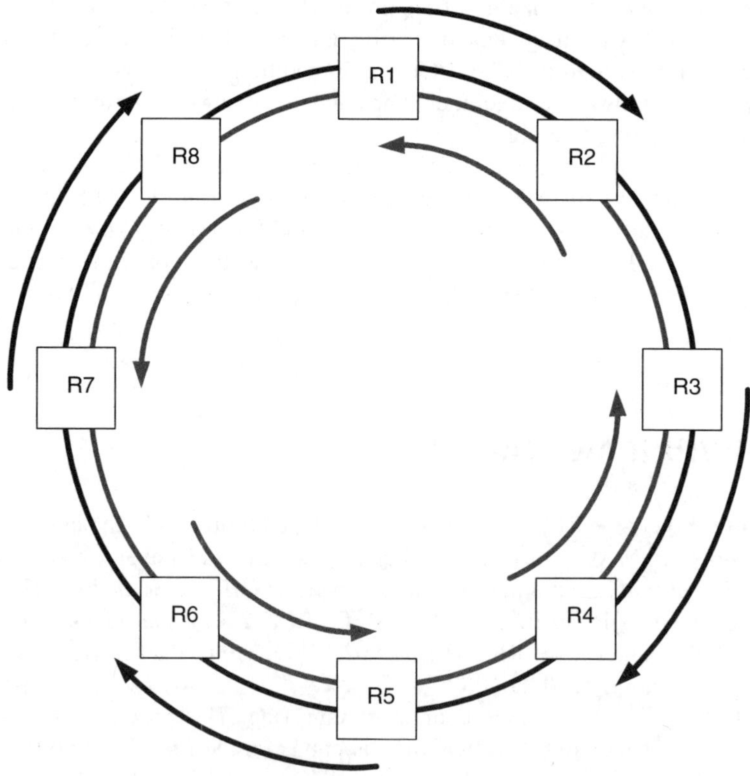

Figure 2.2   Intact SRP dual counter rotating rings

2.5.1   *INTELLIGENT PROTECTION SWITCHING*

One of the features of SRP is the ability to protect against single or double fibre breaks or the total loss of a router. For two such examples, see Figures 2.3 and 2.4. In Figure 2.3 the link between R1 and R8 is broken in both directions (although, in most senses, the same processes would apply if only one of the fibres was broken). As R1 and R8 notice that they are no longer receiving traffic, each will internally wrap the failed transmit circuit back onto the receive circuit. This will create a new topology, and all nodes will be informed that there is a ring wrap. IPS failure protection and restoration are achieved in less than 50 milliseconds, which is equivalent to the restoration speeds required by SONET and SDH systems.

In Figure 2.4, router R1 has failed. In this case, R2 and R8 will identify the failure. Each will then wrap the transmit port normally transmitting towards R1 back to the receive port from R1. Again, the topology will change and all nodes will be informed of the fact that there is a ring wrap on the network.

A well-designed DPT or RPR network can provide exceptionally efficient connectivity over local and wide area networks. In theory, it is possible to build DPT/RPR rings spanning entire continents and including up to 255 nodes. However, this is unlikely to be the most efficient use of this media. Constraining the diameter and, more importantly,

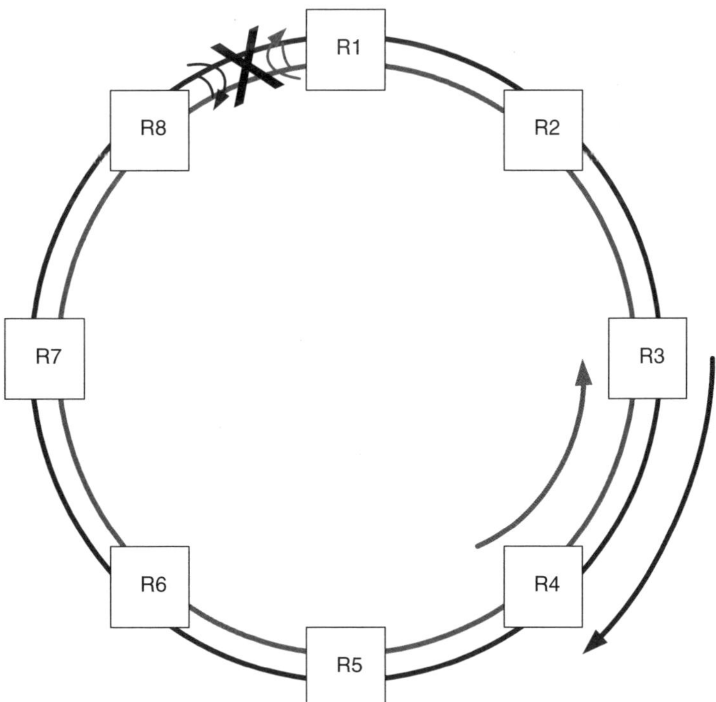

Figure 2.3   Both fibres cut at a single point, ring wrapped at adjacent nodes

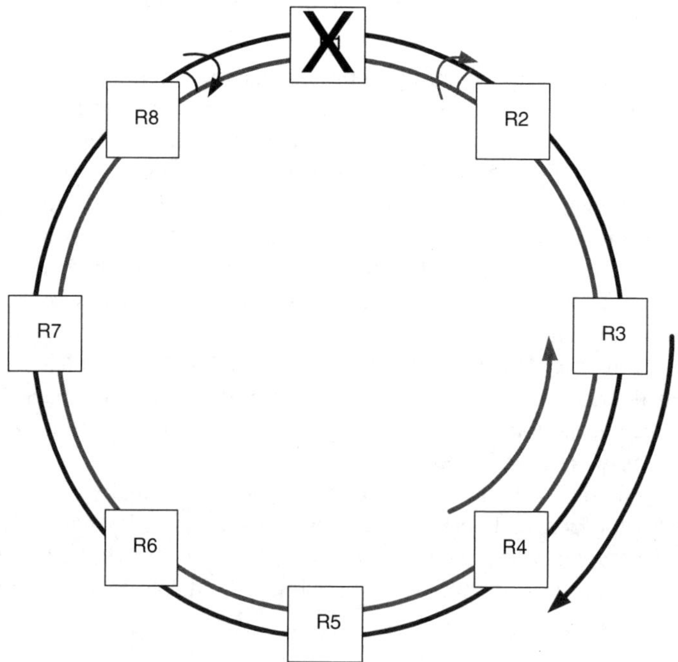

Figure 2.4 Node lost, ring wrapped at adjacent nodes

the number of nodes on the ring will improve the efficiency of the network. In reality, no more than 64 nodes are permitted on a single ring.

2.6 (FRACTIONAL) E1/T1/E3/T3

These lower speed interfaces are commonly used between the provider edge and customer edge. Despite the significant inroads being made by Ethernet to the edge, these interfaces are still dominant because of the lack of widespread availability of Ethernet-based services beyond the urban centres.

The major problem associated with these interfaces is their large number. Traditionally, these are the interfaces over which all customers have been connected to their service providers' networks. When each customer is presented over a single X.21, RJ48, or coax pair, there is a limit to the number of customers who can be serviced in a specific amount of router real estate. In order to overcome this constraint, it is now common for telcos to aggregate many of these circuits over a single SONET or SDH presentation (e.g. 61 E1s presented over a single STM-1, 12 DS-3s presented over a single OC12). This 'channelization' is actually fundamental to the way SONET and SDH work so it requires no additional functionality within those protocols. The only extensions required are to the service offerings of the telcos and to the interfaces on routers.

2.7 WIRELESS TRANSPORT

This is a very broad area, which could fill an entire book of its own. However, in the limited space available, I'll try to cover the basic issues, which any operator of a wireless network will need to address.

The term wireless covers a multitude of transmission techniques and access formats. Wireless transport mechanisms can support point-to-point, point-to-multipoint and broadcast systems. The challenges associated with building wireless networks combine many of the same challenges associated with using any other media along with a whole list of unique problems linked with the air interface. Regulatory constraints, interference, obstructions and even atmospheric conditions all place extra burdens on the network designer. Each of these is discussed below.

2.7.1 REGULATORY CONSTRAINTS

There are a number of different regulatory environments throughout the world, each of which has slightly different rules governing the use of particular frequencies, the maximum transmission power, even the type of service, which may be offered using particular spectrum. Some bands require the payment of licence fees and are more heavily regulated while others are either unlicensed or licence exempt. Even the licence-exempt bands are subject to certain regulations.

2.7.2 INTERFERENCE

Interference refers to any signal that causes the desired signal to be indistinguishable from the surrounding noise. This can be an unwanted signal being received at a very high power level, which masks the required signal being received at a much lower power signal. Alternatively, it can be another source transmitting simultaneously on the same frequency.

2.7.3 OBSTRUCTIONS

The term wireless covers both radio and optical-based systems. Optical systems clearly require line of sight between transmitter and receiver. The line of sight has the diameter equal to the diameter of the aperture of the transmitter. Any obstruction in the path will effectively stop all transmission.

The issue of what constitutes an obstruction is less clear with respect to radio systems. Certain frequencies work better than others without clear line of sight between transmitter and receiver.

2.7.3.1 The Fresnel Zone

It is also important to note the exact implications of the phrase line of sight. Intuitively, you might think that line of sight for radio would be the same as line of sight for an optical system, represented as a tube with a diameter equal to that of the aperture of the transmitting antenna. However, that is certainly not the case.

In fact, there is a rugby ball-shaped (football-shaped for the American readers) zone around the centre line connecting the two antennae. This is called the Fresnel zone. In order to effectively maintain line of sight, it is necessary to ensure that there are no obstructions within 60% of the distance from the centre line to the outside surface of the Fresnel zone. So long as that constraint is met, any calculations can assume clear line of sight.

Some frequencies are particularly susceptible to obstructions; for example, transmissions at 2.4 GHz are exceptionally susceptible to trees in the path. This can be explained by taking a look at the function of microwave ovens. Microwave ovens use this frequency because 2.4 GHz microwaves are particularly well absorbed by water, causing the water molecules to vibrate and get hot. Trees are great big reservoirs of water so it is hardly surprising that they soak up transmissions at 2.4 GHz causing an attenuation of between 0.25 and 0.5 dB per metre.

This is particularly unfortunate because 2.4 GHz is almost universally licence-exempt, making it extremely popular for WLAN systems, as used in 802.11b and 802.11g. Technically, these systems can be used to transmit over distances in excess of 60 km. However, this is severely limited in the presence of trees or other obstructions.

2.7.4   ATMOSPHERIC CONDITIONS

As mentioned above, some frequencies are particularly susceptible to water in the path. Previously, we have discussed this in terms of trees, but the same applies to other sources of water in the path. Rain, fog and snow all cause significant problems for wireless transmission, both radio and optical. For example, rain imposes losses of around 0.05 dB per kilometre on a 2.4 GHz transmission. Even moderately heavy snow or fog will effectively prevent transmission of an optical system beyond a few hundred metres.

These constraints mean that wireless technologies, particularly those in the frequencies vulnerable to absorption by water, are not able to supply high priority links carrying high value traffic without back-up from a more traditional transport media, e.g. fibre or copper-based transmission media.

2.7.5   IF IT IS SO BAD . . .

Given all these issues, it might appear that building a wireless network is not worth the trouble. However, these issues have to be balanced with the ability of wireless networks to reach places where wired services just cannot go, to roll out a network in timescales and at costs within which you could barely plan a wired network.

Wireless is widely seen as a possible solution to the non-availability of broadband services in rural locations. The ability to deliver point-to-point and point-to-multipoint services quickly and cost effectively to areas where there is no cable, where the copper is either of too poor quality or the exchange is too far away to deliver broadband services, is highly attractive to service providers keen to reach those customers and to customers keen to access the broadband services.

3

Router and Network Management

Once your network is operating as expected, it is obviously essential to be able to keep it running smoothly and to know when it stops operating smoothly. Having identified that there is a failure, it is also important to be able to access the network elements affected by and surrounding the fault in order to analyse and repair the failure.

It is also important to be able to correlate events on several devices. Network Time Protocol (NTP) provides a mechanism to synchronize the internal clocks of your network devices. Since the internal clocks are used to timestamp logs, their synchronization is critical to be able to accurately correlate events.

In addition, it is essential to be able to manage the software on your network elements. This is necessary both to add functionality and to overcome faults in previous software.

Similarly, the ability to manage the configuration of your network elements will be critical to the continued smooth operation of your network. Having offline copies of your configurations and a planned regular process for backing up and restoring configurations will allow you to recover more quickly and accurately from a complete failure or loss of a network device.

3.1 THE IMPORTANCE OF AN OUT-OF-BAND (OOB) NETWORK

An out-of-band network is fundamental to the continued operation of a scalable IP network. This is an independent network, which in the event of a failure of part of your network, gives you access to all network devices via an unaffected path. If part of your network

Designing and Developing Scalable IP Networks   G. Davies
© 2004 Guy Davies   ISBN: 0-470-86739-6

fails and you have no OOB network, then you may be unable to reach the network devices affected by the failure to analyse and potentially repair the fault. In that case, you—the network operator—are at the mercy of your telcos. If you are truly fortunate, the fault may 'self-heal'. If you are not, you will be required to send an engineer to the site. This is alright if the failure is in the next city. However, if it is in another country, things get challenging.

OOB networks should ideally provide connectivity to both a management network port and to a console port (if they both exist). In addition, some operators attach a modem connected to the PSTN to the auxiliary port of a device so that even the failure of both the main network and the out of band IP network will not prevent access to the device.

3.1.1 MANAGEMENT ETHERNET

This is generally the most convenient route into the device because it provides a normal IP connection over which you can connect using telnet, SSH, FTP, TFTP, etc. It also provides access for network management and monitoring protocols (e.g. SNMP). If there is no management Ethernet, then the only path over which these protocols can be used is via the in-band (customer traffic carrying) network. Opening up these services on public-facing interfaces leaves the network device more open to attack. Some vendors' devices actually provide the facility to run services separately on the management Ethernet and on the public interfaces. For example, you may want to turn on SNMP, FTP and SSH on the management Ethernet but only enable SSH on the public interfaces. If this facility is available, use it.

3.1.2 CONSOLE PORT

This is generally the last-resort connection into the device. It provides a connection to the management command line interface (CLI), which is completely independent of TCP/IP. This means that even if the routing protocols and the SSH, telnet, SNMP and FTP daemons all fail, and the router is inaccessible via any other path, the console port should still be available. It is also the only place where it will be possible to watch the entire boot process without being disconnected.

In order to manage several devices in a single location via their respective console ports, it is necessary to use a terminal server or console server. Terminal servers generally have a network connection and a modem port. These two ports allow the operator to have easy telnet connectivity to the terminal server via the OOB network. If, for some reason, the OOB network is inoperable, it is possible to connect a modem to the terminal server and to dial in.

3.1.3 AUXILIARY (AUX) PORT

As an alternative to the console port, the auxiliary port provides a mechanism to connect a modem direct to the network device. If all other paths into the device are blocked, this can provide a real last-resort access method to log into a device to find out if it has failed or if it has been isolated from the network by multiple link failures. If there is a terminal server in the hub, then it is probably simpler to connect the modem to the terminal server's aux port.

3.1.4   REMOTE POWER MANAGEMENT

In the worst case, if all else fails, it may be necessary to power cycle the device in order to recover service. If the device happens to be in the same building as your office, then walking through to reboot the device is a simple task. However, if the device is in an unmanned location hundreds of miles from your office, then it is going to be really problematic to recover quickly. One device that can make life a whole lot easier is a remote power device. It permits operators to interrupt and reconnect the power to a network device from a remote location. There are a variety of these devices on the market, access to which is provided via a console port or using the OOB network to connect using telnet, SSH and/or SNMP (see below).

Note: cutting and restarting the power is generally acceptable for network devices (routers, switches, etc.). However, servers (UNIX servers in particular) tend to react badly to being rebooted without a clean shutdown. This is particularly bad because of disk caching mechanisms. By cutting the power without first flushing the caches, you can end up with your disks in an indeterminate state. In the worst case, this can prevent the server from rebooting. This leads to the related issue of uninterruptible power supplies (UPS).

3.1.5   UNINTERRUPTIBLE POWER SUPPLIES (UPS)

The loss of power to a network device is a pretty catastrophic event. Therefore, it is advisable to use a UPS. As the name suggests, a UPS provides a source of power in the event of the failure of the main power supply. Unless you have a room full of batteries, it is unlikely that a UPS will actually be able to sustain power-hungry network devices for more than a few minutes. However, this gives time for an alternative source to kick in (e.g. a generator) without having to reboot each box (with the attendant spike in power draw as all devices power up simultaneously). If there is no alternative power source, the UPS will give the operator time to gracefully shut down servers so that they reboot cleanly.

3.2   NETWORK TIME PROTOCOL (NTP)

NTP is very important to the scalable operation of a network. In particular, it is essential that if an event affecting multiple devices occurs, it is possible to correlate the logs from each of the affected devices to identify simultaneous events. The inability to correlate these events can make it significantly more difficult to identify, isolate and fix any faults. NTP allows individual devices to synchronize their internal clocks so that all network devices report the same time to within a few milliseconds.

NTP works in a hierarchy with stratum one at the top (stratum 0 is 'undefined'). Primary reference servers are stratum one servers. When a server uses time signals from a stratum one server, it becomes a stratum two server. When a server uses time signals from a stratum two server, it becomes a stratum three server, and so on ...

There are many stratum one servers based on a variety of Cesium clocks, GPS clocks, radio time signals, etc. Many of these offer public service so if you do not want to implement your own stratum one servers, it is possible to take service from these servers and have several stratum two servers within your network. All network devices in your network can then take service from each of these servers and their clocks will maintain accuracy as defined by stratum three. It is generally considered to be good practice to only use a public stratum one server if your server is likely to serve a fairly large set of clients and higher stratum servers (100 is considered to be a good figure). It is also considered good practice, if you are going to use a public stratum one server, to operate two or more stratum two servers to service your clients. A good rule of thumb is to operate three servers to provide service to a SP core network and their customers' networks.

In general, the difference between service from a stratum one server and a stratum two server will be negligible, so do not use public stratum one servers unnecessarily.

3.3 LOGGING

Logging is, by definition, the acquisition and storage of large volumes of data regarding the performance of the network. The architecture of the network device defines the mechanisms by which logged information is acquired and the mechanisms and location of the storage of the resulting data. Many modern devices have large volume storage devices built in. These allow the storage of logging and accounting information locally and the transfer of that data in bulk to a central device for post-processing. Those devices with little or no local storage must send all logged information immediately to an external device for storage. This is generally done with syslog and SNMP. It is important to note that syslog and SNMP use UDP as the transport protocol. UDP is inherently unreliable so, during an outage, it can be quite difficult to ensure the delivery of vital messages. This is one of the major benefits of having significant local storage so that logging can be done locally, copied using standard mechanisms to remote servers and, in the event of the failure of standard mechanisms, the locally stored logs can be copied in bulk to the central location.

If you do not log events, it is just impossible to tell exactly what has happened on your network and when. If you offer SLAs, then the inability to identify events, which may impact your SLA, makes you liable to claims from your customers that you have not met your SLA and you have no way to prove or disprove the claim.

3.4 SIMPLE NETWORK MANAGEMENT PROTOCOL (SNMP)

SNMP is a widely used set of protocols for monitoring the health of networks. SNMP is based on four major elements: (1) a network management station; (2) a network management agent; (3) a network management protocol; and (4) management information bases (MIBs).

- The network management station is the workstation to and from which SNMP messages are sent and on which the results are analysed and presented to the operator.

- The network management agent is based on the network device. It responds to requests from the network management station and sends unsolicited notifications of network alert events.

- The network management protocol is the grammar used by the network management station and the network management agent for their conversations.

- The MIB is a dictionary, which describes the objects being given values, the type and range of the values available to each object and an identifier for each of those objects. These objects are formed into a hierarchy of objects.

SNMP provides two operations: GET and SET. GET is used to retrieve information from a network element. SET is used to overwrite a value of the configuration of a network element. The ability to GET and SET values on network devices provides the operator with the ability to both monitor and configure devices remotely using SNMP. The combination of polling (GET) and traps/informs provides a comprehensive mechanism for monitoring all manner of network devices.

3.4.1   SNMPv1, v2c AND v3

There are, currently, three versions of SNMP in widespread use (SNMP v2p and v2u are both obsolete). Essentially, each version provides enhancements to various aspects of the previous versions.

With the standardization of version 3 in RFC 2271 through RFC 2275, the RFCs describing version 1, the initial version, and version 2c, a version with enhanced data types and larger counters, have been made historical.

3.4.1.1 Security in SNMP

SNMP is not an inherently secure protocol. However, security is a significant issue for SNMP, given the nature of the information being passed around. The security (or otherwise) of SNMP can have an impact on the utility of this protocol in larger networks. Since SNMP provides the ability to read and write information to managed devices, it is highly desirable for the manager to be able to decide who may and may not access the information available via SNMP. Versions 1 and 2c use a community string, which is a plain text password. Since each trap and poll sent to/from a network device contains the password in plain text, it is easy to find this password and, with that password, obtain and possibly modify vital network configuration and statistical information. It is not uncommon for providers to leak their SNMP community strings to peers and customers because they try to monitor the links connecting them. This is particularly risky at Internet Exchange Points. If you try to manage the shared network of the exchange, you will transmit your SNMP community string to every single router on the network, regardless of whether they are a peer or not. What is worse, those neighbours may well log the fact that you tried to query their box using an invalid community string.

In order to overcome many of these limitations and associated risks, version 3 of SNMP provides significant improvements in the security of SNMP messages. SNMPv3 provides authentication mechanisms to allow the managed devices and the management systems to confirm the source and integrity of messages they receive. It is also possible for messages to be encrypted so that configuration and statistical information are protected from snooping. In addition, SNMPv3 provides the mechanism to allow different views of the available data. This means that it is possible, for example, for a customer to be given a highly constrained view of the router to which they are connected, while the NOC would be able to view the entire MIB.

It is clear that in an ideal world everyone would use SNMPv3. However, SNMPv3 has not yet been widely implemented. Therefore, both SNMP versions 1 and 2c are still widely deployed in networks throughout the world.

3.5   REMOTE MONITORING (RMON)

RMON is a significant extension to the functionality of SNMP (actually, it's a MIB), that allows the monitoring of a subnetwork as a single unit rather than as a collection of individual interfaces attached to a network device.

3.6   NETWORK MANAGEMENT SYSTEMS

SNMP is a mechanism for the transfer of information relating to the status of your network devices. However, on its own, the raw data is rather dry and can be difficult to interpret. Network management systems are designed to overcome some of these limitations by presenting the information in an easily readable and understandable format. In general, this involves a graphical front end with network diagrams and coloured icons representing nodes and links.

3.6.1   CISCOWORKS

CiscoWorks is a set of network management modules, each of which provides a function within an overall network management system designed to monitor and control Cisco Systems network nodes. The system relies on element managers. These are agents that are responsible for obtaining information from a specific network node type (e.g. a particular

Cisco router model or a particular Cisco switch model) and then passing that information to other parts of the system (e.g. trend managers, provisioning managers, performance monitors, etc.). It is these latter parts of the system to which the network operators have direct access and through which they monitor and influence the behaviour of the managed network nodes.

3.6.2 JUNOScope

JUNOScope is a Juniper Networks proprietary system. It is a graphical user interface for the interrogation of Juniper Networks routers. Currently in very early releases, JUNO-Scope does not provide any configuration functionality.

3.6.3 NON-PROPRIETARY SYSTEMS

In addition to the network management systems developed and marketed directly by the hardware vendors, there are a vast number of non-vendor-specific network management systems from a variety of vendors. Most of these non-vendor-specific systems specialize in specific parts of the overall network management function (e.g. provisioning, red/green monitoring, trend management). For a single vendor to produce a network management system containing all the aspects of an integrated, network management architecture with support for all hardware vendors' equipment would be an almost insurmountable task. This is particularly difficult given the speed at which all the hardware vendors add new features to their software and new hardware modules to their network devices.

3.7 CONFIGURATION MANAGEMENT

As an operator of a network, it is important to keep track of the current configurations on each of your network devices. If a failure occurs, it is essential that you understand the way the device was configured when it was last working. It is also essential, should a network device completely fail, that you know the physical and logical configuration of the device when it was last working. Given how many network failures can be attributed to errors in a configuration file, it is vital that if changes are made to the configuration of a network device, a record is kept of those changes and who made those changes.

3.7.1 CONCURRENT VERSION SYSTEM (CVS)

CVS is a version management system, which is widely used for version control of software. However, it is equally useful in the control of versions of configurations for network devices. CVS allows the administrator to record a base configuration (or software code base) and, as modifications are made, keep track of them as differences from the previous configuration. As changes are made, they are timestamped. Thus, it is possible to recover any version of the configuration that has been saved at any specific time.

It is also possible to create a new branch of the configuration tree, thus creating parallel configurations that can be independently developed. The ability to record updates to a configuration can be restricted to certain users, while a wider list of users can be authorized to pull copies of the latest configurations. This provides a degree of control over who can actually make changes while enabling other network operators to restore the latest configuration to a failed box.

3.7.2 SCRIPTING AND OTHER AUTOMATED CONFIGURATION DISTRIBUTION AND STORAGE MECHANISMS

It is common practice in service providers to use various commercial applications and 'home-grown' scripts to keep records of configurations, update those configurations and also retrieve and analyse network management data.

3.7.2.1 Perl/Expect Examples

Perl and Expect are scripting languages, both of which are well suited to sending information to and parsing the replies from network devices.

Expect is basically a scripting language to pass commands directly to the command line interface of a network device and wait for known responses before passing more commands. It is relatively easy to write an Expect script that uses telnet or SSH to connect to a network device, execute several commands on the command line interface (CLI) and parse the replies. This could be particularly useful for issuing the same set of commands on a list of devices. Table 3.1 is an example Expect script.

Perl, on the other hand, is a much more flexible scripting language for which configuration modifications is just one of a wide variety of possible uses (see Table 3.2). However, control of input and output is somewhat more complex in Perl than it is in Expect. It is not particularly easy to undertake the bidirectional exchange of information required for carrying out management tasks on network devices.

3.7.2.2 CiscoWorks Software

CiscoWorks includes modules specifically designed to keep track of the configurations on Cisco Systems network devices. Cisco Systems devices support the use of telnet, SSH, FTP, TFTP and rcp to access configuration information and these mechanisms are encoded into the CiscoWorks software.

Table 3.1   Expect script to delete MPLS traffic engineering configuration from a Juniper router

Expect script

```
#!/usr/bin/expect -f
#
# This script is used to delete MPLS traffic engineering from a set of
# Juniper routers.
#
# It reads the hostname from a file "listhosts.txt" that contains a list
# of hostnames or ip addresses with one device per line.  It logs into each
# one using telnet.  It then deletes mpls traffic engineering.
#
# To run the script, simply type ./myscript.

set force_conservative 0  ;# set to 1 to force conservative mode even if
                          ;# script wasn't run conservatively originally
if {$force_conservative} {
        set send_slow {1 .1}
        proc send {ignore arg} {
                sleep .1
                exp_send -s -- $arg
        }
}
set timeout 60
# rprompt is the regular prompt of Juniper routers
set rprompt ">"
# cprompt is the configuration mode prompt of Juniper routers
set cprompt "#"
# lprompt is the login prompt
set lprompt "ogin:"
# pprompt is the password prompt
set pprompt "assword:"
# sprompt is the prompt on the server
set sprompt "\[guyd@myserver script\]\$"
# HOME is the base directory for this script
set HOME /home/guyd/script
spawn $env(SHELL)
match_max 100000
expect -ex $sprompt
#Logon onto router and modify the config
set file [open ./listhosts.txt "r"]
while {[gets $file line] != -1} {
    if {[string length $line] != 0 } {
        if { [string first "#" $line] == -1} {
        send -- "telnet $line\r"
        expect -ex $lprompt
        send -- "robot\r"
        expect -ex $pprompt
        log_user 0
```

(continued overleaf)

Table 3.1   (*continued*)

Expect script

```
        send -- "alg0115\r"
        log_user 1
        expect -ex $rprompt
        send -- "configure\r"
        expect -ex $cprompt
        send -- "delete protocols mpls traffic-engineering\r"
        expect -ex $cprompt
        send -- "commit and-quit\r"
        expect -ex $rprompt
        send -- "exit\r"
        expect -ex $sprompt
        }
    }
}
puts "The config change has completed"
close $file
puts [timestamp -format %X-%a-%d-%b]
```

Table 3.2   Perl script to delete MPLS traffic engineering configuration from a Juniper router

Perl script

```
#!/usr/bin/perl
#
# This script is used to delete MPLS traffic engineering from a set of
# Juniper routers.
#
# It reads the hostname from a file "listhosts.txt" that contains a list
# of hostnames or ip addresses with one device per line.  It logs into each
# one using telnet.  It then deletes mpls traffic engineering.
#
# The user must have key based access to the device.  This reduces the need
   for
# interaction to return a password.
#
# To run the script, simply type ./myscript.
#
# Perl doesn't easily support the bidirectional opening of data pipes.
# It is much better at producing formatted text that can be either cut
# and paste into a CLI or used by an API like JUNOScript
#
# Need to work with IPC open2().  Might need Comm.pl or IPC::Chat.

Use IPC::Open2;
Use Symbol;
```

Table 3.2   (*continued*)

Perl script

```
#
# user is the user id for an account with an SSH public key installed on all
   devices
$user = guyd
# rprompt is the regular prompt of Juniper routers
$rprompt = ">";
# cprompt is the configuration mode prompt of Juniper routers
$cprompt = "#";
# HOME is the base directory for this script
$HOME = "/home/guyd/script";
open (HOSTS, "<$HOME/listhosts.txt");
$RDSSH = gensym();
$WRSSH = gensym();
HOST:
while ($hostname = <HOSTS>) {
    $pid = open2 ($RDSSH, $WRSSH, "ssh -l $user $hostname");
SSHLINE:
    while ($line = <STDOUT>) {
        if ($line =  /$rprompt/) {
            print "configure\n" $_;
            print "delete protocols mpls traffic-engineering\n" $_;
            print "commit-and-quit\n" $_;
            print "exit\n" $_;
            last;
        } else {
            next SSHLINE;
        }
    }
    close ($RDSSH);
    close ($WRSSH);
    print "configured $hostname\n";
    next HOST;
}
print "\n\nFinished configuring hosts.\n\n";
close (HOSTS);
```

3.7.2.3 JUNOScript

JUNOScript is a Juniper Networks proprietary, XML-based configuration and device management API. It can be 'driven' using Perl scripts or C compiled programs to create, send,
receive and analyse the XML formatted text returned by the router. JUNOScript can use
SSH, telnet, or SSL to connect to the routing device and exchange information.

3.8   TO UPGRADE OR NOT TO UPGRADE

This decision to upgrade or not is an important one. Upgrading software always brings
with it the risk of a new and unforeseen failure mode. However, it is often the only way

to either fix existing problems or to enable new services or functionality on your network. It is also important to decide whether it is worth upgrading all devices in the network, all devices performing one function in the network or just a single device. Upgrading all devices or all devices performing a particular function in a network will generally involve a significant effort and will carry a significant risk. However, upgrading a single device leaves a network with a variety of software versions. This makes network maintenance, support and troubleshooting more complex.

Before new versions of code are installed on a production network, it is highly advisable to test code in a lab. Clearly, if the code is urgently required to fix an outstanding problem, there will not be an opportunity to pre-test the code. In these circumstances, it is often best to implement a limited deployment in order to limit the risk associated with the deployment of new code.

3.8.1 SOFTWARE RELEASE CYCLES

Every vendor has its own release cycle. Generally, there are several phases in the software lifetime and these can be broadly described as follows:

- *Pre-release*. During this stage, new versions of code are tested, initially within the vendors' own labs (commonly known as alpha) and then within selected customers' labs (commonly known as beta). This code can never be considered bug-free. In general, beta code is somewhat more stable than alpha code. While it is almost never advisable to use alpha code on operators' public networks and beta code is rarely recommended for general use on operators' public networks, there are often occasions where vendors may suggest the use of beta code in isolated cases to solve particular problems or provide particular, urgently needed functionality.

- *Early availability*. Some vendors release code for 'early adopters'. These are users at the forefront of technology with the most advanced requirements. They are usually more willing to tolerate the risk of slightly buggy code in order to get access to the latest features for their network.

- *General availability (GA)*. This is the point at which code is generally considered stable enough for deployment on all networks. The features supported in this code have been deployed widely among early adopters and have been shown to be relatively stable and free of bugs.

- *End of life*. This is the stage in the software life cycle where vendors cease to make the software available to new customers and cease to provide bug fixes for the affected versions. Bugs found in these versions are generally checked and, if present, fixed in subsequent GA releases.

3.9 CAPACITY PLANNING TECHNIQUES

Capacity planning is a widely underrated task by service providers. It is, however, vital to be aware of where your network is reaching capacity and where you need to acquire extra capacity. With provisioning cycles for high capacity links measured in months rather than weeks, it is essential that you understand the traffic trends so that you can place orders for capacity required in several months time.

There are many methods associated with capacity planning. Below are listed some of the more popular capacity planning tools:

- *General traffic graphing tools (e.g. MRTG, RRD).* There are numerous free and shareware applications that provide graphical and tabular analysis of interface throughput statistics. These generally pull statistics from the network device using SNMP, record them in some kind of database and then post-process the results to produce graphical or tabular information regarding the traffic through particular interfaces.

- *Traffic engineering and planning tools.* In addition to the free and shareware applications, which generally provide simple statistics on throughput on an interface, there are numerous commercial applications that provide a much wider functionality. They include applications that take the aggregation of the kind of information used by the 'graphing tools' across an entire network and plot the volumes of traffic flowing on links over time. Often this is used to produce a map of the network with colours used to display the relative 'heat' of a link. Hot links are then flagged for upgrade. The information over time can build up a picture of cyclic traffic flows, which are not visible from the viewpoint of a single interface on a single router. The obvious downside of these systems is that they require the pooling of vast amounts of data to provide any useful results over any period of time.

- *Traps.* It is also possible to use SNMP to send traps when thresholds are broken. This enables alarms to be sent on each event or for a trend to be identified if a particular threshold is regularly broken, for example, at a particular time of day or on a particular day of the week.

4

Network Security

This chapter focuses on network security. It is broken down into categories of physical security and logical access security. There are various elements to each of these including specific security protocols, administrative access control (to prevent unauthorized users from accessing the network elements and wreaking havoc), access tools (to provide secure mechanisms for authorized users to access and recover, view and set information on the network elements), and filters and access control lists (used to prevent unauthorized traffic from traversing the network elements, particularly important in the detection and mitigation of the effects of denial of service (DoS) attacks). Security extends to other aspects (e.g. security of routing protocol messages) but that will be covered elsewhere in the book.

4.1 SECURING ACCESS TO YOUR NETWORK DEVICES

If you are to keep your network running smoothly, then it is essential for you and your customers that you are able to limit access to your network devices to authorized users. By not protecting your network from unauthorized access, you leave yourself open to attack and also provide a conduit for attacks on other networks.

In addition, it is not sufficient to simply permit or deny access. It is also important to be able to classify users into groups with different levels of access (for example, a NOC operator might only need to view information while a network engineer would need to be able to modify configurations, etc.). Finally, it is essential that you are able to identify what has been changed, when and by whom. For this reason (among others), it is critical that each user has an individual account. It is of little value if you know that 'neteng' made a change. Knowing that 'guyd' made the change is much more useful.

Designing and Developing Scalable IP Networks   G. Davies
© 2004 Guy Davies   ISBN: 0-470-86739-6

4.1.1 PHYSICAL SECURITY

Physical security is an often overlooked component of a successful security policy. It is absolutely vital that you strictly control physical access to your network devices. After all, if an intruder has unconstrained physical access to your network, then they will have access to all the buttons and switches on the network devices, the power supplies, the cabling, the interface cards and ports on the router (e.g. the console port and the management Ethernet ports) of all your devices and Ethernet and wireless connectivity to your network resources.

The first task is to ensure that intruders cannot get to the power supplies and the on/off switch. Even if the intruder cannot gain access to your operating network devices, they have the ability to severely disrupt the operation of your network by simply turning off the devices. The ability to power off a device with concurrent access to the console port also provides access on several vendors' devices to mechanisms for resetting the device to factory defaults. This would make it even more time-consuming to recover the box to a functional state.

A carefully chosen, well-protected location for your equipment to which as few individuals as possible have authorized access is the basis on which physical security can be built. Large hubs are often built in commercial secured facilities. A well-run secured facility can provide a high level of security with constrained access to your devices. If this is not sufficient for your purposes, some facilities offer steel cages into which only one customer's equipment is placed. The downside of steel cages is that they can constrain growth within the hub. It can be extremely tricky and can add to the costs due to extra cabling expenses to build a single hub spread throughout several cages in a single facility.

While access via the console port can be authenticated, and in this author's opinion always should be, the console port is often left unprotected. The same premise applies to Ethernet and wireless ports. This is discussed later in this chapter.

If physical security is overlooked, all other aspects of security are irrelevant. Once the physical security is sorted, then you need to take care of the logical security of your network.

4.1.2 AUTHENTICATION, AUTHORIZATION AND ACCOUNTING (AAA)

The first fundamental part of the logical jigsaw is AAA. The authentication and authorization of users as they connect to network devices are vital. If you have no control over who can access your devices and what level of access they have, you are unlikely to be able to keep your network running for more than a few minutes! Similarly, if you are unable to account for who did what and when they did it, you are unlikely to be able to maintain the stability of the network in the long term. Various mechanisms and protocols are available to support authentication, authorization and accounting.

4.1.2.1 Local AAA

This is the simplest mechanism for user authentication. It also inherently provides a significant amount of flexibility. However, it is also by far the most administratively time-consuming and the most difficult with which to maintain the highest levels of security. Each network device contains a database of locally authenticated and authorized users. There is often less granularity of authorization levels and accounting is based on internal logging mechanisms. These authorization levels and accounting mechanisms often tend to be somewhat proprietary in nature and output format thus they are difficult to use in a multi-vendor environment.

The main problems associated with local authentication and authorization are based on the maintenance of consistent details across all network devices. For example, if a user changes their password on one device, the password does not automatically get changed across the network. Also, if a user leaves your company, you have to visit every single network device to remove their authentication. However, this problem can be turned into a benefit if you wish to provide authentication to users for subsets of your network devices. This can be quite difficult if you use a centralized AAA mechanism (see below). Often, this kind of requirement leads operators to develop Perl or Expect scripts to manage the local accounts on all the network devices. This implies that the usernames and passwords are recorded somewhere on a workstation or server, which is likely to impose further security risks. For larger networks, it is invariably better to use a centralized AAA mechanism.

Similar problems exist with local accounting. If all your accounting is local to each network device, then it is not easy to analyse and correlate events occurring on several devices. However, with local accounting, it is obvious to which device the logs relate. With centralized accounting, it is necessary to include some information in the accounting packets to identify the device to which it refers. There are various logging/accounting mechanisms with many being based upon the System Logging (Syslog) protocol. This provides a flexible set of features to log information generated locally and possibly transmit that information to an external server. Syslog can record information at various levels of detail on a scale from 0 (emergency messages relating to a total failure of the network element) through to 7 (debug level messages). It is also possible to record information relating to particular functions of the network device.

4.1.2.2 TACACS/TACACS+

Terminal Access Controller Access Control System (TACACS) and TACACS+ are Cisco proprietary access control systems. Both are centralized AAA systems based on a client server model. TACACS and TACACS+ have been implemented by several other router vendors but, although the details of the protocol are publicly available, the development of the protocol remains entirely under the control of Cisco. In fact, the protocol has remained unchanged for some years and, as far as I know, there are no plans for any modifications to the protocol in the future.

A number of TACACS+ servers are available, ranging from Cisco's own commercial Cisco Secure ACS server to various free implementations. Generally, they are UNIX-based, although Cisco also provides a Microsoft Windows server-based implementation of its ACS server, which supports TACACS+ and RADIUS.

As with all client/server-based AAA mechanisms, the major flaw in the system is related to the loss of the central server. TACACS provides for multiple servers to provide AAA but, if for some reason all servers lose contact with a device, then it is necessary to have a default local login known only to a few carefully chosen engineers.

4.1.2.3 RADIUS

Remote Authentication Dial-In User Service (RADIUS) is an AAA protocol standardized by the IETF. It is almost universally supported by network vendors and provides a mechanism for authenticating users and groups of users and for associating particular access levels with those users or groups of users. It also provides a means of accounting various statistics regarding each user session (time logged in, time logged out, bytes in, bytes out, etc.).

RADIUS is also extensible, providing a means for vendors to create RADIUS attributes specific to their own network equipment's requirements. These attributes are known as Vendor Specific Attributes (VSAs).

There are many implementations of RADIUS servers, some commercial (e.g. Cisco ACS, Funk Steel Belted RADIUS, etc.), some free (e.g. Microsoft IAS, FreeRADIUS). There are implementations that run on various UNIX platforms and Microsoft Windows server platforms, among others.

As with TACACS, it is necessary to create a default local account for use in the event of the failure of the RADIUS servers.

4.1.2.4 DIAMETER

DIAMETER is a framework for authentication, authorization and accounting, the base protocol being defined in RFC 3588. It extends the concepts developed with RADIUS and TACACS+ to provide a comprehensive, reliable, secure, auditable and extensible AAA mechanism.

Specific enhancements over RADIUS include standardized failover mechanisms, transport layer security, reliable delivery, standardized proxy and relay functionality, standardized server-initiated messaging and standardized roaming support. Many of these functions are present in some implementations either as optional standardized functions or as proprietary extensions. However, this leads to difficulty in operating them in a multi-vendor environment.

Applications for DIAMETER are also documented in various Internet drafts. These include Network Access Servers (NAS), Extensible Authentication Protocol (EAP) and Mobile IP.

Currently, there are few widely deployed implementations of elements of the DIAMETER framework, either in network devices or as AAA servers.

4.1.2.5 SSL/TLS

Unlike the previous entries in this section, Secure Sockets Layer (SSL) and Transport Layer Security (TLS) are not mechanisms for the storage and exchange of AAA information. Instead, they provide the means to authenticate yourself to the device with which you are exchanging information and for you to authenticate that device. Authentication is based on the Public Key Infrastructure (PKI). The basis of the security of PKI is that there is a hierarchy of trust. At the top of the hierarchy is a root Certificate Authority (CA), which is considered to be trusted absolutely. There are a number of global root CAs, whose certificates are distributed with many operating systems and applications (particularly web browsers). You can pay a root CA for pre-signed certificates or to get locally generated certificates signed by these global root CAs. However, many enterprises and some service providers run their own, internal CAs, whose certificates are only distributed to hosts that operate within the boundaries of that network. That CA signs certificates for servers, network devices and individuals within the organization exclusively for the purpose of authenticating them to each other. This CA can operate independently of the global hierarchy by simply self-signing its own certificate or it can have its CA certificate signed by one of the global root CAs.

When two entities (individuals, hosts, organizations) wish to authenticate each other, they exchange certificates. When you try to authenticate your peer, you follow the chain of certificates back up the hierarchy of trust from the signed certificate you have received to a point you trust. Generally, this is a chain of just two certificates (i.e. the certificate you receive will have been signed by a root CA).

SSL/TLS is widely used on web servers, and less widely used on mail servers, as a mechanism for authenticating the server to the client. This is most widely used on web servers where the server is being used to take credit card details so that the user can be sure that the server taking their details belongs to the organization to which it purports to belong. In environments where HTML and XML are used to monitor and maintain network devices, SSL/TLS provides a mechanism for both parties to authenticate each other. In most cases, it is tricky for a rogue device to masquerade as a valid device. However, it is possible. It is also eminently possible for a rogue user to masquerade as a valid user.

Running a CA is a difficult task. It can place a particularly heavy administrative load on the network operator if each user is required to authenticate himself using a certificate. Every user would require a personal certificate. These certificates can be difficult to transport between hosts if, for example, the user changes their laptop or if they temporarily need to use a different workstation. However, if this is the type of authentication required, the only alternative is to buy a certificate for each user from a global root CA. Even with bulk discounts, this could be prohibitively expensive. As usual, you are left with a compromise. You could internally manage a CA, with the attendant administrative load and costs or pay a third party for the numerous individual certificates from a CA. As always, you must make your own compromises between cost, complexity and security.

4.2   SECURING ACCESS TO THE NETWORK INFRASTRUCTURE

As discussed in the introduction to physical security, in many networks, particularly enterprise networks, the ability to access the network infrastructure can be limited effectively by controlling the ability to access the holes in the wall inside the premises of the network operator. If there are few routes into the network (again, particularly common in enterprise networks), then it is also easier to apply filters or access lists that deny access from external devices to any network device on your network. Traditionally, service providers were cautious about implementing access lists on the ingress to their network because of the performance impact this had on all traffic. However, as discussed in Chapter 1, modern ASIC and network processor-based designs are no longer vulnerable to these limitations.

4.2.1   AUTHENTICATION OF USERS, HOSTS AND SERVERS

Despite the importance and benefits of physical security of your premises, there are certain situations where physical protection is not sufficient. In these cases, it is necessary to authenticate the hosts or individuals connecting to the network before full access to network resources is granted. It may also be desirable/necessary for network access devices to authenticate themselves to the users logging in. The authentication of users, hosts and network devices is particularly important for wireless networks, where the physical boundaries of the network operator's premises are unlikely to contain the transport media entirely. It is also possible for someone to install a rogue network device outside (or inside) the building and use it to obtain authentication information from subscribers. Various aspects of authentication are described below. While wireless is used widely as an example, the elements are, in general, equally applicable to wired connections.

4.2.1.1 802.1X

802.1X was initially designed as a mechanism for authenticating hosts and users as they connected to infrastructure switch ports (e.g. switch ports on an Ethernet switch). It limits the access to the network infrastructure to the minimum required to authenticate the host or user until authentication is complete. 802.1X was based on the Extensible Authentication Protocol (EAP), created by the IETF, but was subsequently extended to provide support for authentication of hosts and users requesting connections to wireless infrastructure. 802.1X is now extremely widely implemented by both Ethernet and wireless equipment vendors.

It is important to note that 802.1X is only a framework for authentication. It provides no encryption functionality in and of itself.

4.2.2   ENCRYPTION OF INFORMATION

In addition to the need for authentication, depending upon the nature of the transport media, it might be desirable to apply encryption to the packets being transported. Again, this is particularly important in wireless environments, where it is possible for someone to sit outside your building and simply listen to all the traffic flying by. In an unencrypted environment, all your company's confidential data could be flying by, being intercepted by any third party.

4.2.2.1 WEP, WPA, TKIP and 802.11i

Several protocols have been created over time to solve the problem of maintaining the privacy of data transported over wireless infrastructure. The first, Wired Equivalence Privacy (WEP), is often used to perform both authentication and encryption functions. However, it has been shown to suffer from several severe weaknesses and, even with key lengths of 128 bits, on a moderately busy network can be routinely compromised in a matter of hours. In order to overcome some of these weaknesses, a mechanism for automatically changing the keys used by two peers has been devised. This mechanism does not overcome the inherent weaknesses. It merely works around them making it more difficult to compromise a key before it is discarded in favour of a new key.

Given the inherent limitations of WEP, work was begun on the standardization of a new extension to the 802.11 group of standards, 802.11i, which would provide enhanced encryption services to 802.11-based wireless networks. It is based on a combination of 802.1X, EAP, the Advanced Encryption Standard (AES) and the Robust Security Network (RSN). RSN is designed to dynamically negotiate authentication and encryption functions between the client and the access point. Work on this standard has been 'steady' but, at the time of writing, it is still not complete. Implementation of everything in the 802.11i standard is likely to require changes not only to software but also to existing, deployed hardware if performance is not to be unacceptably poor. However, despite the incomplete status of the 802.11i standard, some vendors are starting to introduce hardware capable of supporting the proposed standard.

In order to provide enhanced security in the period until 802.11i is standardized and compatible hardware widely deployed, the WiFi Alliance (an industry body) devised the WiFi Protected Access (WPA). WPA is a framework of enhanced security protocols and algorithms designed to work on currently deployed hardware. Part of the WPA framework is the Temporal Key Integrity Protocol (TKIP), which provides a mechanism for changing the shared key used by the client and network access device based on the passing of time or packets and for exchanging the updated keying information.

4.2.3   ACCESS TOOLS AND PROTOCOLS

Having AAA mechanisms is all well and good, but if the access tools and protocols are not secure, then all the AAA mechanisms are for nothing.

4.2.3.1 rsh, rlogin, rexec and rcp

Remote shell (rsh), remote login (rlogin), remote execute (rexec) and remote copy (rcp) are all access mechanisms that were developed as part of the UNIX networking suite.

They provide the facility to login to, execute commands on and copy files to and from remote devices. They have been implemented on some network devices. However, serious concerns about the security of these protocols, particularly the ability for authentication on one device to automatically imply authentication on all devices and the passing of user credentials in the clear, have led many network operators to explicitly disable these protocols on any device, which supports them. In addition, the implementations of these protocols have traditionally been the source of many security flaws in many operating systems. Many network device vendors now do not implement these protocols on their devices because of the inherent security problems. Even when they are implemented by the vendors, most SPs disable them.

4.2.3.2 SSH/SCP

Secure Shell (SSH) and Secure Copy (SCP) are secure alternatives to rsh, rlogin, rexec and rcp. Due to their enhanced security features, SSH and SCP are widely favoured over their poorly secured predecessors by SPs. You can log in to the CLI of a network device and copy files to and from a network device respectively. SSH works on the basis of a public and private key pair for the server and for the client. During the initial exchange, the server passes its public key to the client. The client can either be prompted for a username and password or, alternatively, a public key associated with an individual user can be preplaced onto the server. Once the server and the client both have each other's public keys, they use those keys to encrypt all messages for the other device. It is only possible to decrypt the messages encrypted with the public key if you are in possession of the private key. Thus, once the client has encrypted a message using the server's public key, only the server can decrypt using its private key (which is, of course, kept absolutely secure on the server).

4.2.3.3 Telnet

Telnet is a simple, unsecured terminal access application. It is by far the most widely implemented terminal access program but is widely disabled by network operators because of its inherent lack of security. The main issue with telnet is that all the user's credentials are passed between the client and the server in the clear. This means that any user attached to any network in the data path can snoop the traffic on that network and view the username and password.

4.2.3.4 FTP

File Transfer Protocol (FTP) is a simple and relatively efficient mechanism for transferring files to or from a network device. However, it is not secure since all user credentials are passed from the client to the server in the clear. It is based on TCP so any missed packets will be retransmitted based on mechanisms inherent in TCP. FTP is a widely implemented protocol, many implementations of which have had significant security flaws. For this reason, if SCP is available, it is almost invariably preferred over FTP. If SCP is not available, then it is common practice to disable the FTP service on any device except for the period during which it is actually being used.

4.2.3.5 TFTP

Trivial File Transfer Protocol (TFTP) is, as the name suggests, a more trivial protocol for the transfer of data. It is based on UDP, so there is no inherent acknowledgement and retransmission of packets. Therefore, it is necessary for TFTP to implement that functionality. TFTP is insecure and generally requires no authentication at all. For this reason, it is often turned off by operators if an alternative file transfer mechanism exists. If no alternative is available, it is advisable to protect the service either with a secured wrapper (a proxy application that provides access control mechanisms to applications that do not possess that facility) or with access control lists.

4.2.4   IP SECURITY (IPSEC)

IPsec is a suite of protocols, which were initially devised as an integral part of the IPv6 suite of protocols. However, they have been extended to support IPv4. The original framework RFC 1825 has been made obsolete by RFC 2401. It provides both encryption and authentication of data. Numerous encryption and authentication algorithms are supported within the framework of IPsec.

At the time of writing, IPsec is not used for access to devices or the transfer of information to and from those devices.

IPsec provides two functions: authentication and encryption. The two functions are based on Authentication Headers (AH) and Encapsulating Security Payload (ESP) respectively. These are IP protocols 51 and 50 respectively.

The basis for IPsec is called the security association (SA). An SA is a unidirectional connection over which secured traffic is carried. If a connection between two hosts requires both AH and ESP protection then at least two SAs must be established. (see Figure 4.1).

Thus, in order to protect a normal bidirectional exchange between two hosts, at least two SAs are required, but such an exchange may require several parallel SAs in each direction. Each SA can be uniquely described by the three parameters: Security Parameter Identifier (SPI), a destination address and the security protocol identifier (i.e. ESP or AH).

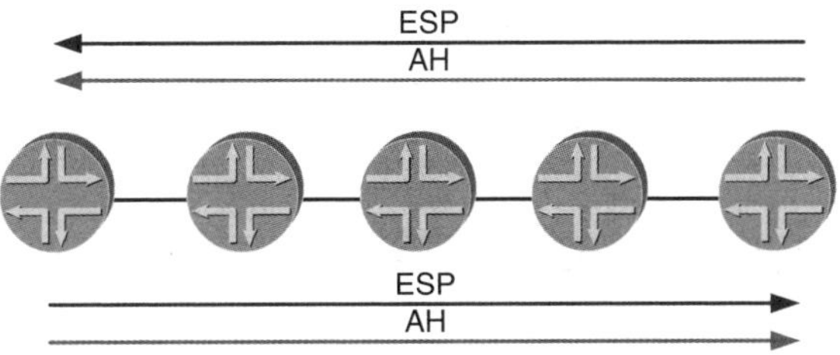

Figure 4.1   Security Associations unidirectional

There are two modes in which IPsec operates; transport mode and tunnel mode. Transport mode operates exclusively between two hosts. Since the IPsec header is located between the IP header and the Layer 4 header (TCP, UDP, etc.), with ESP in transport mode, only the Layer 4 header and beyond are protected. With AH, protection is extended into part of the IP header.

Tunnel mode is effectively a secured IP tunnel. If either end of the SA is a gateway, the SA must be operated in tunnel mode. The only exception to this is if the ultimate destination of the data is the gateway, in which case the gateway is operating as a host. In this case, it is permissible to operate the SA in transport mode. With tunnel mode there are two IP headers. The IPsec header is located between the first and second IP header and affords protection to the entire tunneled IP packet irrespective of whether the SA is an ESP SA or an AH SA.

IPsec supports a variety of user authentication mechanisms ranging from simple username and password pairs through to full, PKI-based certificate exchange.

IPsec is widely used to provide VPN functionality both between two fixed devices (e.g. branch office to central office) and between mobile devices and a central fixed device (e.g. remote workers connecting back to the office). It is entirely capable of providing secure connections between two mobile devices. However, this is not currently a widely used application of IPsec. In order to have the necessary trust to set up and maintain an SA between two mobile devices, it would be almost essential to use a certificate exchange to authenticate the two devices/users to each other.

Scaling IPsec is a serious challenge. As mentioned previously, it is possible for host-to-host IPsec SAs to be established so that all communication between any arbitrary pair of hosts is protected. However, the number of SAs this would require on each host would present a computational burden on the hosts. In addition, in order to make host-to-host IPsec a realistic option, it would be necessary to have a PKI-based authentication system. The exchange and maintenance of certificates between each pair of hosts would add to the overheads associated with any host-to-host exchange. However, assuming that the exchange between the hosts is non-trivial, the extra overheads will not be significant. The administrative overhead of obtaining, signing, maintaining and distributing all the certificates may well be a very significant extra overhead in a host-to-host model. So, if we look at the alternative, where one or both ends of the SA are gateways, then we have the problem that, as the number of remote locations to which any individual gateway must connect grows, in order to sustain the SAs required to protect communications, there is a significant amount of state to maintain. The maintenance of state is always a constraint upon scaling. The maintenance of all these SAs involves a significant amount of memory and processing. Since the gateway is effectively a throttle point, it is also necessary for these devices to have significant packet forwarding capability.

4.2.5 ACCESS CONTROL LISTS

An Access Control List, in its most generic sense, is any mechanism that provides a means to filter packets or routing information based on any of the attributes of the routing or routed protocols. This is an almost universally implemented mechanism; however, the syntax, functionality and flexibility of the various implementations differ widely.

Unfortunately, in some devices, particularly those with a software-based routing and forwarding function, the application of access control lists can cause a dramatic performance impact on the router. Generally, if the access control lists are not implemented in hardware, all packets have to be analysed in the CPU. This starves the CPU of cycles for other tasks (route calculation, running the CLI, etc.).

In JUNOS software, access lists are implemented using routing policy and firewall filters.

4.2.5.1 Preventing and Rate Limiting Access at the Ingress to the Network

The first line of defence for your network is the edge routers. Intuitively, it is clear that this is the best place to apply filters to protect your network. By preventing 'spurious' traffic from ever entering your network, there is inherently less risk (note: there is still some risk) to each of your individual network devices. It is arguable that in applying filters at the edge of the network, you remove the need for any filters on all non-edge devices. I feel that taking it to this extreme would be foolhardy since there are still threats originating within your network (e.g. unauthorized users, disgruntled employees) against which filters can provide effective protection.

4.2.5.2 Preventing and Rate Limiting Access to Certain Hosts/Networks

This is one of the most common uses for access control lists. Basic access control mechanisms allow the filtering (either outright blocking or rate limiting) of packets destined for or sourced from particular hosts or networks.

4.2.5.3 Preventing and Rate Limiting Access to Particular Protocols/Processes

In addition to filtering all packets either destined to or sourced from particular hosts or networks, it is possible to further refine filters based on the protocol and/or port of the source/destination.

4.2.6   RFC 1918 ADDRESSES

RFC 1918 describes three ranges of IP addresses that are designated for private use within non-Internet connected networks. I am sometimes asked if using RFC 1918 addresses within the core of your network will assist in the protection of your core devices. In theory, RFC 1918 addresses should not be advertised outside any autonomous system (nor accepted by any neighbour), so it should not be possible for any device outside your network to reach a core device numbered from RFC 1918 address space. However, it is possible to force the routing of packets via a particular edge device on your network from which the RFC 1918 addressed host is reachable. This is known as source routing and should always be disabled on all routers. There is little justification for enabling source routing anywhere in an operator's network.

It is also important to note that this provides no protection against the misbehaviour of your own customers. A significant number of your customers will have a default route towards your network so they will be able to reach all your RFC 1918 addressed nodes (and your nodes will be able to reply).

In addition, when communicating with a host, many client applications attempt to resolve the address of the host to a name using the domain name system (DNS). In general, most people have not created entries on their local name servers for the resolution of all RFC 1918 addresses to a dummy name. Since they have no local entry, the name servers are forced to attempt to resolve the addresses through the standard DNS processes. Thus, the root servers will receive vast numbers of queries from clients trying to resolve RFC 1918 addresses. This places an exceptional load upon the root name servers. In fact, the operators of some of the root name servers have resorted to placing a firewall in front of the name servers to block all requests for resolution of RFC 1918 addresses.

Therefore, when asked, I advise people that the use of RFC 1918 addresses in your core provides scant protection for your network and places an unreasonable load upon critical shared network infrastructure. Don't do it!

4.2.7 PREVENTING AND TRACING DENIAL OF SERVICE (DOS) ATTACKS

Denial-of-service attacks are a growing problem for which there appears to be no single long-term solution. Putting in place measures to prevent and mitigate the effects of DoS attacks and to trace the source of the attack can mean the difference between providing the service which your customers expect and failing to meet your SLAs.

The wide variety of potential DoS attacks means that a number of strategies are required to combat them. In some cases, simple filtering can stop the attacks. In others, the effects of the attack can only be limited by rate limiting particular traffic types or patterns. In general, DoS attacks fall into several categories:

- Sending packets, either malformed or in such quantities, that cause the CPU of a device to become overwhelmed and incapable of performing other tasks.

- Sending malformed packets, which are incorrectly handled by the operating system causing the system to either crash or provide uncontrolled access to unauthorized users with which those users can disrupt service.

- Sending packets in such volume that links are over-subscribed, preventing the transport of valid customer traffic.

Over subscribing a CPU is less of an issue with modern routers in which the routing module is separated from the packet forwarding module. Since fewer packets are, by default, sent to the routing module, it is harder to select a packet that can be subverted but that still gets transferred to the routing module. Also, it is quite common to rate limit packets through the link from the packet forwarding module to the routing module. This protection is normally left on to mitigate the effects of a full-scale attack. In the event of an ongoing attack, it is essential that filtering and rate-limiting be applied at the borders of your network.

Sending malformed packets that cause software to fail is an attack that can normally only be overcome by patching the software. The effects of some of this type of attack can be mitigated by filtering or rate-limiting particular types or patterns of traffic. However, this can impact valid traffic so it is often preferable to use this kind of measure only as a temporary measure to limit the effect until a patch can be applied to the software.

Sending packets in large volumes is often achieved by creating a distributed attack. An attacker places a piece of code on a very large number of poorly protected hosts. Each of those hosts is then remotely configured (or may be configured as part of the malicious code) to attack a particular host or network. Although each host may only be able to contribute a few tens or hundreds of kilobits per second, if hundreds or thousands of these vulnerable hosts are coordinated to start transmitting simultaneously, the effect can be quite catastrophic at the target network. The best defence against this kind of attack is to apply filters at the borders of your network and leave them in place indefinitely. If this is not possible due to the architecture of your network devices, then it is important to have the filters configured and ready to be applied at very short notice.

4.3   PROTECTING YOUR OWN AND OTHERS' NETWORK DEVICES

In addition to the protection of your network devices, it is important to be able to make use of your network devices to protect other devices that are reached via your network (e.g. your customers' networks, managed servers, etc.). In many cases, the same functionality described above in relation to access to your network devices can be used to provide controlled access to resources via your network. The benefits and constraints associated with each of the protocols when used to access your network devices apply just as much to access via your network.

5

Routing Protocols

This is one of the major areas over which the network architect has a great deal of influence. Decisions made by the network architect can dramatically impact the behaviour of the IP traffic as the network grows and the number and type of services increase. Poor decisions at the design stage can severely impact your ability to grow and could necessitate a complete redesign of the architecture, causing major inconvenience to your customers.

Routing, as a generic term, is the process of selecting the next hop via which a destination is best reached. Best, in this case, is a rather nebulous concept, usually associated with the shortest path measured in some arbitrary units. The idea of the best route can be modified by routing policy. The term routing also incorporates the concept of forwarding. Forwarding is the actual process of receiving a packet on one interface and then transmitting the packet out of an interface based on the destination address.

The next hop to any destination in the network can be explicitly configured on any router. This is known as static routing. Alternatively, routers can use dynamic routing protocols. Routing protocols are the means by which network layer reachability information (NLRI) is exchanged between two or more routers. This exchange is essential if packets are to be forwarded correctly to their destinations. It would be possible to statically configure all routing information on all routers throughout the entire Internet but, in reality, this would be totally impractical since static routing cannot route around failure and it would be administratively intolerable to have to manage static routing in a large network. Static routing has a valid place in the Internet but it certainly cannot replace dynamic routing protocols.

Routing protocols can be categorized in several ways, including link state vs. distance vector, intra-autonomous system[1] vs. inter-autonomous system, classful vs. classless.

---

[1] An autonomous system is a contiguous collection of networks and nodes under the control of a single administrative authority.

Designing and Developing Scalable IP Networks   G. Davies
© 2004 Guy Davies   ISBN: 0-470-86739-6

Below we will take a quick look at a number of routing protocols widely used in service provider networks and at the mechanisms each uses to improve scalability. Not all protocols neatly fit into these categories, as there are some derivatives and hybrids. However, they provide useful categorizations on which to base this discussion.

5.1 WHY DIFFERENT ROUTING PROTOCOLS?

Many people, particularly those new to the field, ask why we need different routing protocols for our internal routes and our external routes. There are several reasons but the most compelling is that each type of protocol is optimized for its particular purpose. In this respect, there are two main types of routing protocols: Interior Gateway Protocols and Exterior Gateway Protocols.

5.2 INTERIOR GATEWAY PROTOCOLS (IGP)

IGPs are optimized for smaller sets of prefixes within an autonomous system but provide rapid convergence. Generally, it is wise to place fewer, more stable prefixes in your IGP. Ideally, this should be constrained to the links, which actually form part of your own network (i.e. links and loopback[2] interfaces attached your core routers, edge routers and, possibly, between edge routers and hosts/servers). IGPs do not usually scale well when comparatively unstable prefixes or large numbers of prefixes are carried. Some feel that it is best not to use an IGP to carry any routes associated with individual customers at all (i.e. point-to-point links and customers' own networks). However, others feel that the IGP is fundamentally a mechanism for distributing reachability information for BGP next hops. If this is the case, then the point-to-point links could reasonably be considered as a valid IGP route. In the end, this is strictly a policy decision for the network operator. Those holding the former viewpoint feel that the IGP databases should be kept as small as possible and that it is unnecessary, given the ability to set BGP next-hop self,[3] to carry any information relating to networks other than those in the core. They point out that if the IGP is unable to stabilize and converge, then it is of no value. Thus, it is essential to make the task of the IGP the bare minimum.

Making a specific statement about how many routes an IGP can hold before it becomes a problem is not easy. The actual number of routes depends upon a huge number of variables, e.g. the design of the protocol, the implementation of that design by the vendor, the way the protocol has been configured by the operator, the stability of the various links, the processing power of the routers, the ability of the routers to transfer routing

---

[2] Loopback interfaces are virtual interfaces that are not associated with any physical interface. Such interfaces are always up and reachable as long as any one of the physical interfaces on the router is up.

[3] Next-hop self is explained in more detail later in this chapter.

information from the forwarding plane to the control plane, and so on. In an Internet in which the global routing table is around 150 000 prefixes, it is preferable to keep the number of prefixes in your IGP in the hundreds or low thousands. In the tens of thousands, most IGPs (irrespective of how well implemented, configured and sustained by the routers they are) will become unstable. As mentioned above, an unstable IGP is a disaster, because it not only affects your internal routing but also your ability to reach any external destinations.

Today, SPs generally use OSPF or IS-IS, the larger ones tending more towards IS-IS.[4] Other IGPs are described here for completeness and because they have a significant role to play in enterprise networks, even moderately large ones.

5.2.1 OPEN SHORTEST PATH FIRST (OSPF)

OSPF is a classless, link-state IGP developed in the late 1980s and early 1990s by the IETF. It distributes routing information within a single autonomous system. OSPF floods routing information throughout the entire domain using link state advertisements (LSAs). Each router runs a shortest path first (or Dijkstra) algorithm to derive a complete network topology. All routers in the network have an identical copy of the link state database (LSDB), however, each views the network as a tree with itself as the root of the tree.

OSPF supports a hierarchical model in which 'areas' are created. Areas are groups of routers and links. The border between areas falls on routers known as Area Border Routers (ABRs). Each area is identified by a 32-bit identifier (often presented in the form of an 'IP address', although it is totally unlinked to any IP address). The core of a network is area zero and all non-zero areas must connect to area zero. All traffic from one non-zero area to another non-zero area must pass through area zero.

ABRs also support the concept of route summarization. Route summarization is yet another compromise. It reduces the amount of routing information stored, which eases the burden on resources such as memory and processing power. However, route summarization can lead to the selection of sub-optimal paths. This is particularly true where more than one ABR provides connectivity between area zero and a single non-zero area.

While OSPF within a single area operates in a link state mode, it operates in a distance vector mode between areas. When routing information is passed between areas, the information regarding the actual source can be lost. Instead, the ABR becomes the advertising router of the summary and it becomes the 'vector' towards which the traffic should be sent. This is fundamentally due to the summarization of routing information from one non-zero area upon entry into area zero. This summary information is sourced by the ABR and the topology of the non-zero area is hidden. This information is then flooded throughout the rest of the network.

One side effect of this distance vector behaviour is that some form of loop detection or loop avoidance mechanism is required. The chosen mechanism is to constrain the network to a star-like model. By mandating that all non-zero areas must be connected only to the

---

[4] The reasons for this choice are discussed in more detail later in this chapter.

backbone area, it is possible to remove the risk of any loops. Figure 5.1 demonstrates a valid network architecture within OSPF, while Figure 5.2 demonstrates an invalid network architecture within OSPF.

In order to bypass the constraint that all areas must be directly connected to area zero, a mechanism called virtual links was devised. This allows a non-zero area to be connected to area zero via another non-zero area (see Figure 5.3).

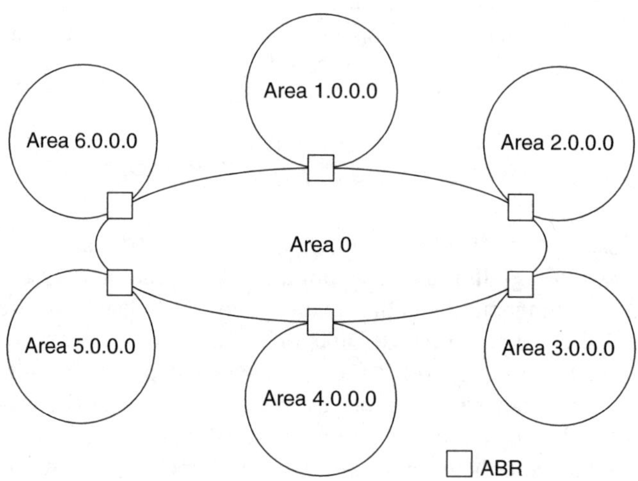

Figure 5.1 Multi-area OSPF hierarchy

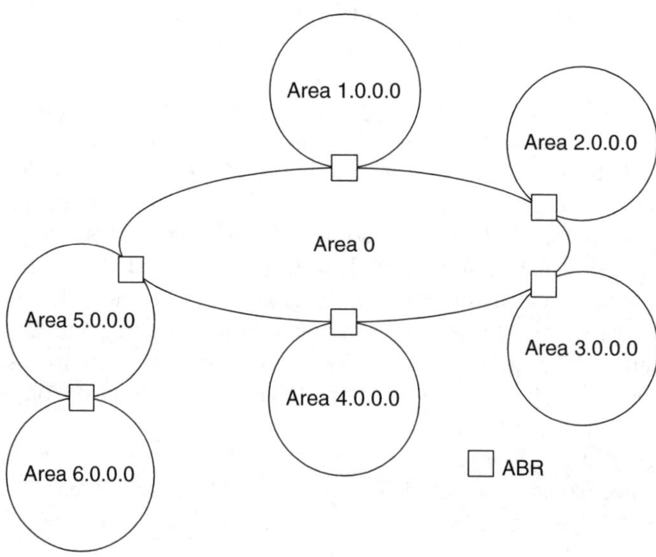

Figure 5.2 Illegal OSPF hierarchy

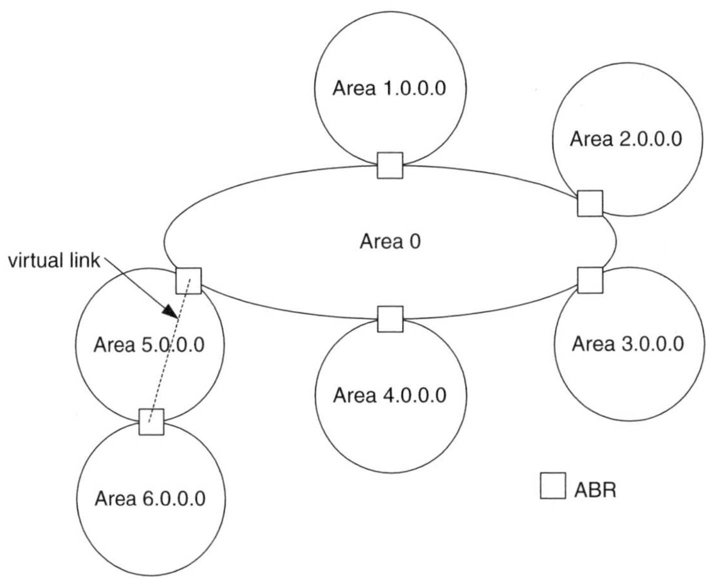

Figure 5.3 A virtual link makes this hierarchy legal

5.2.2 AUTHENTICATION OF OSPF

OSPF is vulnerable to the insertion of spurious data, either maliciously or inadvertently. Authentication is one mechanism often used to protect against the insertion of invalid data into OSPF. OSPF provides three states of authentication: none, simple authentication and MD5 authentication. It is required, if you configure authentication, that you configure the same authentication method throughout a single area. It is also required in RFC 2328 that authentication be configurable on a per-interface basis. If you cannot configure different authentication mechanisms on a per-interface basis, then it follows that the same authentication type must be configured throughout the entire autonomous system.

- *None*. As the name suggests in this mode, OSPF exchanges are actually not authenticated at all. This is generally the default configuration of OSPF implementations. It provides no protection whatsoever.

- *Simple password authentication*. OSPF provides the ability to use a clear text password. Both neighbours must use an identical clear text password in order for the neighbour relationship to be fully established. Since the key is passed in the clear with every authenticated packet, it is a relatively simple matter to snoop the key and inject spurious routing updates. Therefore, this is not a mechanism which would be recommended for a large network.

- *MD5 authentication*. OSPF is able to use MD5 authentication mechanisms. Each neighbour must have the same MD5 key for any exchange of OSPF packets to occur. A checksum is created for each packet and that checksum is checked using the MD5 key defined on the receiving router.

MD5 is not immune to attack. There are some well-known and well-documented mechanisms for attacking authenticated exchanges, which rely on the ability to snoop the link between the neighbours. Since OSPF is generally used within a single IGP, then it is important to ensure physical and logical security of your routers (see Chapter 4) so that your authenticated OSPF exchanges cannot be snooped.

OSPF supports the creation of neighbour relationships over multiple hops (e.g. virtual links). This feature makes OSPF particularly vulnerable to the injection of spurious information, even from outside your AS. Therefore, it is advisable to deploy one of the available authentication mechanisms. Since clear text passwords can be sniffed and replayed, if the feature is available, it is recommended that you deploy MD5 authentication with OSPF.

5.2.3 STUB AREAS, NOT SO STUBBY AREAS (NSSA) AND TOTALLY STUBBY AREAS

The fundamental reason for the creation of each of these non-standard area types is scalability. If you have edge routers with only limited resources, it may be necessary to constrain the number of LSAs reaching those routers, particularly if you have very large numbers of LSAs in your IGP. Each presents a compromise between total routing knowledge, optimal routing and the associated extra requirements on memory and processor or the reduction in routing knowledge, reduced resource requirements and potential routing inefficiencies. As with many of the issues raised in this book, it is all about compromise.

- *Stub areas.* Stub areas permit area-local Router LSAs and Network LSAs and require summary LSAs covering all inter-area routes. All other LSA types are not permitted in stub areas, therefore; it will be necessary for the ABRs to generate a default summary LSA in the stub area if full reachability is to be maintained within the autonomous system. This default route can be the single prefix, which subsumes all the inter-area routes, thus reducing further the total number of prefixes in the area.

- *NSSA.* NSSAs are identical to stub areas with the exception that they also permit locally generated AS external LSAs. A special LSA is required for this purpose. The NSSA External LSA is effectively identical to an AS External LSA but requires a different LSA type so that it can be differentiated from AS Externals generated elsewhere in the autonomous system. NSSA External LSAs are converted, before transmission out of the NSSA, into AS External LSAs by one of the ABRs. This allows the injection of prefixes from other routing protocols, a feature expressly forbidden with normal stub areas.

- *Totally stubby areas.* Totally stubby areas take the concept of stubbiness to the extreme. The only permitted LSAs are area local router and network LSAs plus a summary LSA for a default route from each of the ABRs. This reduces the LSAs in the LSDBs of all non-ABRs to the absolute minimum to maintain complete reachability within the autonomous system. However, it also takes the risk of sub-optimal routing to its maximum.

It is difficult to make any specific recommendation with respect to which of the above mentioned non-standard area types to implement without specific details of the architecture and hardware in use in the specific network. However, some useful guidelines can be presented. In general, it is desirable to have the greatest degree of routing knowledge throughout the largest possible part of the network. Compromises on this issue should only be made in circumstances where the hardware in the area is incapable of sustaining the entire LSDB. The extent of the compromise should depend entirely upon the extent of the limitation of the hardware. Wherever possible, such compromises should be kept to a minimum so that optimal routing can be maintained.

5.2.4   OSPF GRACEFUL RESTART

When a router restarts, it affects not just those routers directly connected to the restarting router. The change in topology is flooded to every router in the entire area and each is required to run an SPF calculation. It is also possible, depending upon the nature and location of the topology change, that this is also propagated beyond the boundaries of the area in which the topology changes occur so the impact can be significant.

In addition, in modern routers in which the forwarding and routing processes are separated, it may be that the forwarding function is unaffected by the failure of the routing function. In this case, rather than simply tearing down all the neighbour relationships and clearing all associated forwarding state, it seems appropriate to maintain the forwarding tables in exactly the same state they were at the time of the failure. This allows the router to continue forwarding traffic, despite the fact that the routing function is inoperable.

There are several constraints on the ability of a router to restart gracefully:

- If, during the restart process, there is any other change in topology, then the restarting router will exit graceful restart mode.

- If the restarting router is already waiting for another router to restart, it will not do a graceful restart. This is related to the previous item because this represents another topology change, which causes the first restarting router to exit graceful restart mode.

- When a router restarts, it informs its neighbours how long they should expect to wait before it is back. If the restarting router has not recovered before the timer expires, then the restarting router exits graceful restart mode.

- Graceful restart is an optional function. Therefore, it is necessary for neighbours to negotiate its use at the time the neighbours establish a relationship. If one neighbour does not support graceful restart, then graceful restart cannot be used by that restarting router. In addition, at the time the router begins the restart process, it starts sending grace LSAs. If these are not acknowledged, then graceful restart is unlikely to be supported by the neighbour that did not acknowledge. In this case, the restarting router does not use graceful restart.

5.2.5   OSPFv3

OSPFv2, the current version of OSPF, is designed specifically to carry IPv4 NLRI. In order to support IPv6 NLRI, OSPFv3 has been developed. OSPFv3 does not support IPv4 NLRI. OSPFv3 provides support for IPv6 for all the capabilities supported by OSPFv2 for IPv4. However, some of the features are implemented in different ways. For example, authentication for OSPFv3 is provided by the AH and ESP functionality inherent in IPv6 rather than providing a mechanism within the OSPFv3 protocol.

5.2.6   INTERMEDIATE SYSTEM TO INTERMEDIATE SYSTEM (IS-IS)

IS-IS is a classless, link state IGP originally designed by the International Organization for Standardization (ISO) to carry Connectionless Network Protocol (CLNP) routing information. Each node running IS-IS is known as an intermediate system (IS). This is the OSI term for a router. It is still widely used as a routing protocol on many Data Control Networks (DCN) for transmission networks, since they traditionally supported OSI protocols rather than IP as the transport over which the management systems were run.

However, at around the same time as OSPF was being developed by the IETF, IS-IS was proposed as a routing protocol for IPv4. IS-IS is inherently extensible so it was a relatively simple matter to add the necessary Type/Length/Value tuples (TLVs) to accept IPv4 prefixes (and subsequently to carry IPv6 prefixes).

In common with OSPF, IS-IS distributes routing information within a single autonomous system. It also supports a hierarchical model. Unlike OSPF, however, IS-IS provides two levels of link with no special identification of a core area (i.e. no specific area zero) and no identification to differentiate one non-core area from another. Each IS is either a level 1 IS, which routes traffic entirely within its own area, or a level 2 IS, which is capable of routing traffic between areas. If a level 1 IS wants to send traffic to another area, it forwards it towards the nearest level 2 IS. An area containing only level 2 ISes is analogous to OSPF area zero. All areas containing level 1 and level 2 ISes are somewhat analogous, but not quite identical, to NSSA areas in OSPF. Each link can be either in level 1, level 2 or both. Also, unlike with OSPF, the border between areas falls on a link rather than an IS, or router.

IS-IS is widely used by very large service providers. The reasons for this choice are, to a certain extent, due to the circumstances at the time at which those networks started growing rapidly. In the early to mid-1990s, these service providers' networks started to grow to the point where existing routing protocols (primarily IGRP and EIGRP) started to reach the limit of their scaling. Those service providers started looking for an alternative IGP and were presented with two choices (OSPF and IS-IS). At the time, almost all of those service providers used equipment from Cisco Systems and Cisco's implementation of OSPF was somewhat less stable than its implementation of IS-IS. Given that stability was one of the main reasons for the change to a new IGP, this pointed convincingly towards IS-IS. Probably under pressure from some of its larger customers, when resources were limited, Cisco tended to develop new features in IS-IS before OSPF. Over time, Cisco's implementations of both IS-IS and OSPF have stabilized and there is now no significant difference between the two in terms of functionality or stability. Despite this, if you were to ask Cisco which protocol they recommend for use in a large SP network, they would invariably advise you to use IS-IS.

5.2.7   AUTHENTICATION OF IS-IS

As with OSPF, IS-IS is vulnerable to the insertion of invalid routing information. However, since IS-IS is carried directly over the Layer 2 protocols, IS-IS is fundamentally less vulnerable to the injection of spurious data than those routing protocols, which run over IP because in an Internet environment, Connectionless Network Service (CLNS) is not routed. However, the availability of the feature provides us with an interesting choice to make. Generally, it is worth applying authentication mechanisms if they are available. The implementation of authentication might add minimally to the computational overhead associated with the protocol. However, the overhead is small and the processors in modern routers so powerful that this is not a significant issue.

IS-IS provides the same three authentication modes as OSPF: none, simple authentication and MD5 authentication:

- *None*. In this mode, IS-IS protocol exchanges are not authenticated. There is no protection in this mode. This is the default configuration on most, if not all, implementations of IS-IS. Since other, effective authentication mechanisms are available, it is not recommended to leave IS-IS unauthenticated unless your chosen vendor has not implemented any authentication mechanisms for IS-IS.

- *Simple authentication*. This method uses a plain-text pre-shared key, which is passed with the authenticated protocol data units (PDU) to the neighbour. Anyone with the access to snoop traffic between two adjacent nodes will see the password and be able, therefore, to inject spurious routing updates. For this reason alone, simple authentication is not recommended unless no better alternative exists.

- *MD5 authentication*. Because IS-IS does not run over TCP and does not support the creation of neighbour relationships between nodes, which are not directly adjacent on physical links, the difficulty of exploiting this vulnerability is somewhat greater. Despite this, IS-IS also supports the use of MD5 authentication between adjacent nodes.

  The availability of MD5 authentication is as good a reason as is necessary to use it. It does not make your routing exchanges invulnerable to attack but it significantly reduces the risk.

- *Authentication of hello exchanges*. In addition to the authentication of messages in an established adjacency, it is also possible to independently enable authentication of the

IS-IS Hello PDUs (IIH) through which the adjacency is initially established. If IS-IS authentication is enabled, hello authentication is automatically enabled.

If hello authentication is available, it is likely that all IS-IS PDUs can be authenticated using MD5. In this case, it is sensible to authenticate all IS-IS protocol exchanges.

5.2.8   IS-IS GRACEFUL RESTART

As with OSPF, when an IS-IS router (an IS) restarts, it causes SPF calculations to be performed on all routers within its area or, in the case of a level 2 router, throughout the entire domain. In a relatively unstable network, this can cause a significant load on the processor calculating the routing table.

In common with graceful restart for OSPF, graceful restart for IS-IS is optional. Unlike OSPF, there is no negotiation of the capability between the neighbours. The restarting router merely includes a restart TLV with the RR bit set in the IIH PDUs. This informs the neighbour that it should not re-initialize the adjacency. It also requests a complete sequence number PDU (CSNP) from any neighbours with adjacencies via point-to-point links and requests that any neighbour sets the SRM flags (used to indicate that an LSP needs to be refreshed) for all of the LSPs learned from the restarting router.

If a router is operating an interface on a shared network in both level 1 and level 2, then it must send and receive IIH PDUs for each level with the RRs set. Over a point-to-point link, a single IIH PDU can be sent for both levels. Irrespective of the type of network, separate CSNPs must be sent for each level. IIH PDUs are sent until the exchange of CSNPs is complete, which ensures that the restarting router is completely updated by all its neighbours following the restart.

5.2.9   ROUTING INFORMATION PROTOCOL (RIP)

RIP is a distance vector routing protocol. There are currently two versions of RIP. Version 1 is classful. It assumes the legacy classful prefix lengths (0.0.0.0 through 127.0.0.0 are all /8, 128.0.0.0 through 191.255.0.0 are all /16 and 192.0.0.0 through 223.255.255.0 are all /24). Version 2 provides extensions to support classless routing. It associates an arbitrary prefix length between 0 and 32 with each network address.

Being a distance vector protocol, RIP suffers from the 'counting to infinity' problem. The 'counting to infinity' problem is based on the fact that an unreachable route is always given a metric of 'infinity'. It is patently not possible to actually have a metric which is equal to infinity, so some absolute value has to be assigned to be equal to infinity. In RIP, this value is 16, which constrains RIP networks to a diameter of 15 hops. If a path is longer than 15 hops, it is effectively equal to infinity and therefore unreachable.

RIP's default timers also mean that it is exceptionally slow to converge. Each node announces its complete routing table to its neighbours every 30 seconds so, in the worst possible case, a change experienced on one side of a maximum diameter network would take seven minutes to propagate to the other side of the network. In addition, routes learned from a neighbour will, by default, not expire for 180 seconds. If a router fails on a switched network, shortly after announcing a set of prefixes, it will take up to 180 seconds before the neighbouring routers even register the loss of the neighbour and its routes. Only then will it start propagating information to its neighbours saying that the prefixes are no longer reachable via that path.

Using RIP for anything other than one or two hop networks is, in reality, too slow to converge. However, there are still a few peripheral products in use on service providers' networks, which support no dynamic routing protocols other than RIP. In this situation, the only alternatives are RIP or static routing. It is often preferable to use static routing in this case rather than use RIP.

5.2.10   *INTERIOR GATEWAY ROUTING PROTOCOL (IGRP) AND ENHANCED INTERIOR GATEWAY ROUTING PROTOCOL (EIGRP)*

IGRP and EIGRP are Cisco proprietary routing protocols. Although the details of the protocol behaviour have been published, no other vendor may implement IGRP or EIGRP. Therefore, their use is constrained to Cisco-only networks.

5.2.10.1 IGRP Specifics

Like RIP, IGRP is a classful, distance vector routing protocol. It operates in a very similar way to RIP, periodically broadcasting its entire routing table via all interfaces. It also requests information from all neighbours when it starts up to obtain a complete routing table as quickly as possible.

IGRP uses the concept of autonomous systems as process identifiers. IGRP autonomous systems do not necessarily correspond to the concept of autonomous systems in OSPF or BGP. In IGRP an autonomous system represents a set of routers using a common IGRP routing process to exchange information. It is possible, within a single IGP, to define multiple IGRP instances between which it is possible to tightly control the exchange of routing information.

Unlike other routing protocols, IGRP uses a composite metric system that allows it to overcome, to a certain extent, the constraints of RIP. IGRP still uses the hop count to a destination as an element of the metric and can be used in networks up to 255 hops in diameter. Composite metrics are derived from four link characteristics: bandwidth, delay, load and reliability. By default, IGRP uses just the bandwidth and delay, only using load and reliability if specifically configured to do so.

Bandwidth and delay are static values not directly linked to the actual bandwidth and delay of the associated link. By default, each link type has a particular value. However, these values can be manually configured. Bandwidth is transmitted as a three octet number derived from the bandwidth in kilobits per second scaled by a factor of 10^7. Delay is also used to represent unreachability. Delay is measured in tens of milliseconds and transmitted as a three octet number. An unreachable destination has a delay of 0xFFFFFE (approximately 167 seconds). The total metric is calculated as the minimum bandwidth plus the sum of the delays, end to end.

The importance of bandwidth and delay in deriving the IGRP metric means that it is essential that they are chosen carefully, accurately and consistently across the entire network. Changes to either bandwidth or delay should be made with extreme care.

In addition to the extended network diameter, IGRP has some other benefits over RIP. Primary among these is the ability to do unequal cost load sharing. Unequal cost load sharing uses a value called variance. This is simply a multiplier that indicates a range within which a higher metric must fall in order for it to be considered for use in unequal cost load sharing. If the higher metric is less than or equal to variance multiplied by the lower metric, then it is a feasible route.

Like RIP, IGRP is easy to configure. It is this simplicity that made it attractive to operators of networks running exclusively Cisco equipment.

5.2.10.2 EIGRP Specifics

Despite its name, EIGRP is not really a direct derivative of IGRP. It is Cisco proprietary but it is classless and is neither a pure distance vector nor a link state protocol. The routing algorithms employed by EIGRP, described in more detail below, incorporate some features of distance vector and some features of link-state algorithms.

EIGRP does provide compatibility with IGRP and uses the same concept of an autonomous system representing a collection of routers using the same routing process. In fact, EIGRP uses the same algorithm as IGRP scaled by 256 to provide greater granularity in the metric values.

There are mechanisms for NLRI to be exchanged automatically between EIGRP and IGRP. In fact, it is (almost) as simple as assigning the same autonomous system number to both protocols on a router. The main constraint is that being classful, IGRP is unable to carry any information relating to the length of a particular prefix and simply assumes that it has learned an entire classful prefix. If care is not taken, this can lead to routing loops and black holes.

EIGRP uses the Reliable Transport Protocol (RTP) to manage the delivery of EIGRP protocol packets. Packets are multicast to the well-known multicast group 224.0.0.10 using a Cisco proprietary mechanism known as reliable multicast. Responses to multicast packets are unicast. Two sequence numbers are used to ensure ordered delivery of packets. Each transmitted packet contains the sequence number from the source that increments with each new transmitted packet and the sequence number of the last packet received from the destination.

Unlike IGRP or RIP, which use the periodicity of updates to identify the loss of an adjacent router and its associated routes, EIGRP does not have periodic updates. Therefore, it is necessary to use a neighbour discovery and maintenance mechanism to keep track of neighbours. EIGRP uses periodic Hello packets to maintain its neighbours. EIGRP Hello packets also inform the neighbours how long they should wait without hearing any Hellos before declaring the sending node dead and updating the routing information.

Unlike a pure distance vector protocol in which a single shortest route via one next hop router is chosen and all other routing information relating to that destination is discarded,

EIGRP maintains a complete topology of the network. This is somewhat similar to a link-state protocol and provides incredible benefits with respect to the speed of convergence. Rather than a router having to wait for invalid routes to expire and for discarded routing information to be retransmitted by alternative next hops, EIGRP provides a mechanism for an alternative path to be identified in advance of any failure and, if no such alternative path exists, a router can query its neighbours to find alternatives.

Having described how much EIGRP resembles a link-state protocol, it is necessary to point out that EIGRP still transmits routes as distances and next hops (vectors) to directly connected neighbours. However, unlike IGRP and RIP, EIGRP only sends updates when a routing change occurs, only sends the changed routing information and only sends those updates to other routers impacted by the changed routes.

Particularly in enterprise networks, where bandwidth in wide area networks is generally more restricted and for whom bandwidth is usually more expensive, this reduction in protocol traffic can be a major benefit, particularly when the network is moderately unstable.

As stated earlier, EIGRP does not use a pure link-state algorithm. Link-state algorithms rely on each router to individually calculate the topology of the network from flooded routing information. EIGRP, on the other hand, uses a distributed mechanism whereby calculations are made in a coordinated way throughout the network. This is known as diffusing computations.

5.2.11   DIFFUSING UPDATE ALGORITHM (DUAL)

DUAL is the algorithm through which updates for the diffusing computations are exchanged. In order for DUAL to function, it is essential that the following conditions are met:

- Neighbour state is recognized and recorded within a fixed time.

- Updates are received complete and in the correct order within a fixed time.

- Updates are processed in the order in which they are received within a fixed time.

DUAL uses the term feasible distance (FD) to identify the lowest calculated metric to a destination. The feasibility condition (FC) is met when a neighbour advertises a distance to a destination that is lower than the FD for that destination. If the FC is met, then the advertising neighbour becomes the feasible successor (i.e. the next hop). The feasible successor is always closer to the destination and cannot have a path to the destination back via the router with the FD. This makes it impossible to create a routing loop.

For every destination, for which at least one valid feasible successor exists, the following information is stored:

- the FD to the destination;

- all feasible successors;

- the advertised distance from each feasible successor;

- the distance via each feasible successor, as calculated locally based on the advertised distance and the cost of the link to the feasible successor;

- the interface via which the feasible successor is reached.

Of all the feasible successors, the one via which the lowest metric has been calculated is declared the successor. The information associated with the successor is used to populate the routing table.

When a current successor advertises the loss of a prefix, the router looks for another feasible successor. If one is found, then the normal process to select the successor from the list of feasible successors is carried out. If no feasible successors are found, then the route is placed in Active mode and queries are sent out to trigger the diffusing computation. The route remains in Active mode until replies are received to all queries.

5.2.12 STUCK-IN-ACTIVE

So what happens if a reply is never received to a particular query? If an adjacent router is unable to reply (maybe it has gone down since the last hello), then the route would remain in Active state forever. However, when a route is placed in Active, an active timer is started and decremented towards zero. If the timer expires before a reply is received, the neighbour is declared dead. All routes associated with that neighbour are removed from the routing table. In addition, all routes from the dead neighbour are assigned an infinite metric so that DUAL is able to correctly deal with the loss of the neighbour.

This situation has led to significant problems in the past. In growing provider networks, particularly those with comparatively low bandwidth links or links into single routers of vastly differing capacities, the problems occurred when there were topology changes within the network resulting in lots of queries. The topology changes resulted in certain links being overwhelmed by traffic and thus the routers were unable to respond to the incoming queries. This, in conjunction with the network instability, caused lots of routes to be stuck in active. This caused EIGRP sessions to be dropped, which in turn caused large numbers of routes to be withdrawn. And so the problems cascaded through the network.

These problems can be mitigated by careful application of route summarization or by the creation of routes to null0 (the discard interface) for the affected prefixes.

5.2.13 WHY USE EIGRP?

This all sounds horribly complex. However, the complexity is mostly hidden from the user. EIGRP, like IGRP, is relatively trivial to implement and operate. This makes it particularly attractive to small network operators who do not necessarily have hordes of highly skilled network engineers to keep the network ticking along. It can be pretty much turned on and left alone to get on with it.

5.3 **EXTERIOR PROTOCOLS**

Exterior protocols are necessary for the exchange of routing information between autonomous systems. It would be impractical to use an IGP for this purpose for a number of reasons. As mentioned earlier, IGPs are optimized for small sets of prefixes. Exterior protocols, on the contrary, are optimized for the exchange and manipulation of very large sets of prefixes.

5.3.1 BORDER GATEWAY PROTOCOL (BGP)

BGP is the only exterior gateway protocol in widespread use in the Internet, today. The current version of BGP is version 4. It is safe to say that every multihomed network (not just multiply connected but truly multihomed to different upstream providers) needs to run BGP in order to provide any dynamic resilience to their network.

BGP runs over the Transmission Control Protocol (TCP), which means that it is based on a peering system whereby two routers maintain a TCP connection over which all BGP messages are exchanged. Any single router may be maintaining many such sessions at any one time.

As an exterior gateway protocol, BGP is used for exchange of routing information between autonomous systems, which are groups of routers and the links between them controlled by a single administrative authority. Each autonomous system requires a globally unique identifier, which is a 16-bit integer (0 to 65535, 0 is reserved, 64512 to 65535 are available for private use and, since they should not be announced into the Internet, do not need to be globally unique).

BGP is a path vector protocol. This is not a true distance vector protocol but has significant similarities. The path to the destination is described as a list of these identifiers representing each of the autonomous systems back to the originating autonomous system in the order in which they must be traversed. Every time a prefix is advertised to another autonomous system, the router adds its own AS number to this list, the AS_PATH. Thus, the AS_PATH is created as the announcement of the prefix passes from the destination towards the source. Each AS number in the AS_PATH is analogous to a 'hop' with a distance of one in a distance vector protocol.

In addition to providing details of the 'distance' to a particular prefix, the AS_PATH also performs another function. Upon receiving a prefix from a neighbour, a BGP speaker checks to see if its own AS number is in the AS_PATH. If it is, then this prefix would create a routing loop. Therefore, such prefixes are rejected.

Although BGP is fundamentally designed to exchange routing information between autonomous systems, it is still necessary to use BGP to exchange routing information between routers in the same autonomous system. The former configuration is known as external BGP (eBGP) and the latter is known as internal BGP (iBGP). While the underlying protocol is fundamentally the same, there are some important differences in the way it is operated and some different constraints on each.

5.3.1.1 External BGP (eBGP)

BGP is fundamentally a protocol designed for the exchange of inter-autonomous system NLRI. External BGP includes all BGP sessions between devices in your own autonomous system and those in any other autonomous system. This includes your upstream transit providers (potentially including other parts of your own company operated as separate autonomous systems), your peers and your customers. There is no inherent difference between the mechanisms used to exchange routes in these different types of eBGP session. However, each type of eBGP session has different aims and needs to be treated differently. Policy mechanisms are used to achieve this. Policy is discussed fully in Chapter 6. However, below is a brief description of the aims and requirements of each type of eBGP session:

- *Transit providers.* This is a session over which you learn all Internet routes (or a significant subset of those routes, e.g. all European routes). It is usually treated as the route of last resort, since you will almost inevitably be paying for traffic going over the link to your transit providers.

- *Peers.* This is a session over which two organizations exchange their own routes and those of their customers at 'no cost'[5] to each other. The cost of connecting to each other is usually shared equally. Routes learned via this type of session are likely to be preferred over those learned from a transit provider since there is little or no cost[6] associated with traffic to peers.

- *Customers.* This is a session to a customer, which pays you to carry their traffic. Generally, you provide either full Internet routes or a subset of Internet routes and a default route. In effect, this makes you their transit provider. Routes learned from customers are generally preferred over all other sources because customers pay you for carrying traffic.

5.3.1.2 Internal BGP (iBGP)

If you use more than one router to connect to external BGP neighbours, then those routers must be connected by a contiguous path of BGP-enabled routers. In the most basic configuration, all routers running BGP in a network must peer with all other routers running BGP in the same network.

This constraint is necessary because no prefixes learned from an iBGP neighbour may be announced to another iBGP neighbour. This is a measure to prevent routing loops. This additional constraint is required because a router cannot add its own AS number

---

[5] There is usually no fee paid by one party to the other for the transport of their customers' data.
[6] There is usually some cost associated with joining the Internet exchange point or the provision of the circuit between the peers, even if it is simply the cost of a Gigabit Ethernet port on both parties' routers and the fibre between them. However, compared to the cost of transit service, this is likely to be negligible.

to the AS_PATH when sending prefixes to an iBGP neighbour. If it were to do so, the neighbour would detect its own AS in the AS_PATH and reject the prefix.

The full mesh of internal BGP sessions is the basis of the n^2 problem. In fact, using the basic configuration, you will have $n(n-1)$ peering sessions configured, where n is the total number of BGP speakers in the network. Thus, when you add the $n+1$th BGP router to your network, you actually must reconfigure every single BGP router in your network (n devices). As your network grows, this can become a huge administrative burden. Two mechanisms have been developed to reduce this burden: route reflectors and confederations.

5.3.1.3 Route Reflectors

Route reflection is a mechanism which allows network operators to configure only partial meshes of routers while maintaining full connectivity. The mechanism relies upon special BGP routers called route reflectors. These special devices are permitted to break some of the basic rules of internal BGP. Specifically, the route reflectors may announce routes learned from one iBGP neighbour to another iBGP neighbour (as mentioned above, this is forbidden in basic iBGP). In order to restore the protection against routing loops, extra BGP attributes have been created. Those attributes are ORIGINATOR and CLUSTER_ID. A router announcing a prefix into a BGP autonomous system sets the ORIGINATOR to its own router-id. Each route reflection cluster has a CLUSTER_ID on the route reflector, which is added to the CLUSTER_ID_LIST as the prefix crosses the boundaries of BGP clusters.

All route reflectors must be fully meshed within a 'core'. It is permissible to either fully mesh all clients in a particular route reflection cluster or to instruct the route reflector to reflect announcements from one client to another within the same cluster.

The choice of whether to run clients in a full mesh or use client-to-client reflection raises some interesting issues. Intuitively, the benefits of client-to-client reflection are pretty clear. This certainly places the lowest administrative load on the network operator. Adding a new client simply involves configuring the client and the reflectors for that cluster. However, client-to-client reflection can be a contributory factor in the problem described as persistent route oscillation. This problem is more fully described in Chapter 6 and in great detail in RFC 3345.

Clearly, a full mesh inside the cluster imposes its own n^2 scaling issues as each of your clusters grows. If a cluster grows unwieldy, it is perfectly valid to create clusters within clusters, thus creating a hierarchy of route reflection (see Figure 5.4).

It is common practice to run route reflectors in pairs. When a pair of route reflectors service a single cluster, it is not obligatory for those route reflectors to use the same CLUSTER_ID. There are valid arguments for both options. If you use the same CLUSTER_ID on both route reflectors, then you reduce the total amount of state held on each router right across the network. This extra routing information will inherently impose extra load upon the processors, causes more protocol exchanges between iBGP neighbours and requires more memory. However, depending upon the way in which iBGP has been implemented by the operator, it is possible that the extra state may provide extra

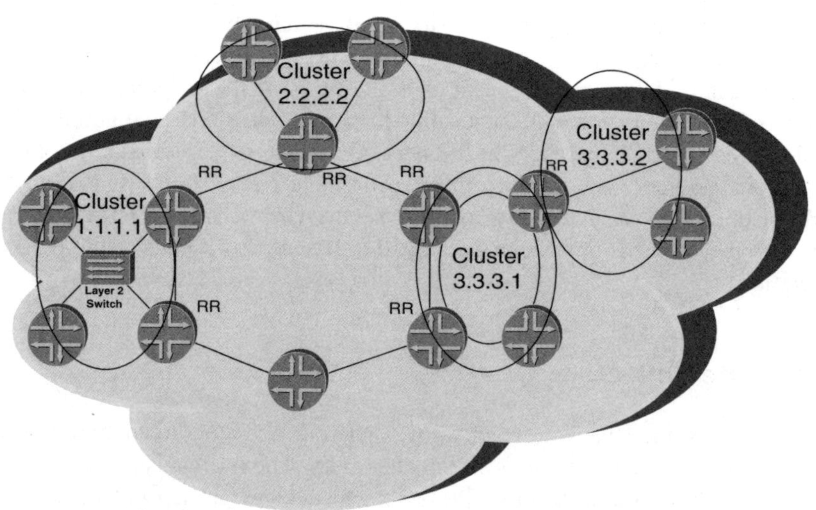

Figure 5.4 Route reflection

resilience. For example, if the alternate path is already available to the route reflector clients upon failure of the preferred path, then the time taken to restore traffic flow would be slightly reduced.

5.3.1.4 Confederations

Using BGP confederations is another mechanism for reducing the full mesh of iBGP sessions. It works on the premise of dividing a large autonomous system into a set of smaller sub-autonomous systems (confederation members), which are each interconnected in a limited and controlled fashion. Every router within any single confederation member (or sub-AS) must be either fully meshed or must employ route reflection mechanisms as described above.

Between sub-ASes, routers talk eiBGP (also known as confederation eBGP). As the name suggests, it is a bit of eBGP and a bit of iBGP. There are some differences between eiBGP and both eBGP and iBGP. For example, LOCAL_PREFERENCE is exchanged by iBGP neighbours but not by eBGP neighbours. LOCAL_PREFERENCE is also exchanged by eiBGP neighbours.

Conversely, the Multi-Exit Discriminator (MED) is used by eBGP neighbours to decide which path to use, but not by iBGP neighbours. MEDs are used by eiBGP neighbours for the same purpose. Also, when announcing a prefix to a different confederation member via an eiBGP session, the BGP speaker will add his own sub-AS number to the AS_PATH (in a special object known as an AS_CONFED_SEQUENCE), just as is done on eBGP sessions. This provides the same level of loop prevention as is provided by the AS_PATH in traditional ASes.

When announcing a prefix to an AS that is not a member of the confederation, the entire AS_CONFED_SEQUENCE is purged from the AS_PATH and the confederation's

AS number is added. Thus, despite all the sub-ASes being individually visible within the confederation, the whole confederation still appears as a single AS to any external neighbour. So a single peer network could peer with routers in two different confederation members but the peer network would perceive that it was peering in two locations with the same AS.

To demonstrate this concept, in Figure 5.5, AS1 has five confederation members (65000, 65001, 65002, 65003 and 65004). A prefix with the AS_PATH 10_11_12 is learned via an edge device in sub-AS 65004. It is announced to sub-AS 65000 the AS_PATH (65004)_10_11_12.[7] Sub-AS 65000 then forwards the prefix to sub-ASes 65001, 65002 and 65003 with the AS_PATH (65000_65004)_10_11_12. Sub-ASes 65001 and 65002 now forward the prefix to AS2. Rather than forwarding 1_(65001_65000_65004)_10_11_12 and 1_(65002_65000_65004)_10_11_12 respectively, they both strip the AS_CONFED_ SEQUENCE from the AS_PATH before prepending the confederation's AS. Thus, both send 1_10_11_12 as the AS_PATH.

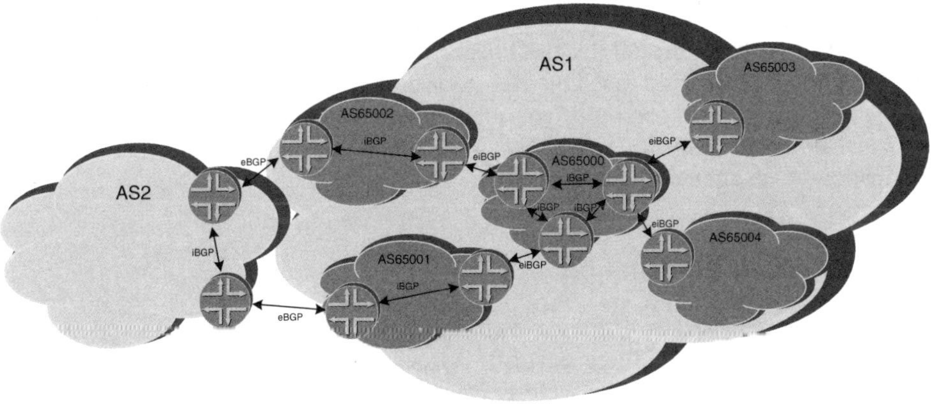

Figure 5.5 BGP confederations

5.3.2 AUTHENTICATION OF BGP

Perhaps more than any other protocol, BGP has the potential to be the means of injection of spurious routing information. It also has the capacity for the widest impact. This has been amply demonstrated by the catastrophic events when a router in Florida was subject to a software fault causing deaggregation of CIDR blocks. This routing information was injected back into the global routing tables and caused major problems for significant parts of the Internet. This event caused problems for several hours, even after the network devices at the source of the problem had been completely isolated from the Internet! This demonstrates the degree of vulnerability of BGP to invalid information, although it has

---

[7] The list of numbers in brackets represents an AS_CONFED_SEQUENCE.

to be stressed that current authentication mechanisms could not have limited the effects of these faults because the prefixes were being learned from a 'trusted' neighbour.

However, similar effects could be produced by a rogue device injecting bogus routing information into unprotected peers on a shared network. Given this vulnerability, it is possibly more important than with any other routing protocol that we authenticate our BGP neighbours.

The authentication mechanism is not actually part of the BGP protocol but is part of the underlying transport protocol. Since TCP does not provide any simple authentication mechanism, the only standardized mechanism for authenticating BGP sessions is MD5 signatures.

5.3.2.1 MD5 Authentication

BGP runs over TCP and can, therefore, use the MD5 authentication mechanisms for TCP messages to authenticate all BGP protocol exchanges. This provides a method to authenticate messages as coming from a trusted source. Given the extent to which the whole Internet can be affected by the injection of spurious information into the global routing table, it is certainly recommended to use this mechanism to ensure that you are peering with the device you think you are peering with. It is important to realize that MD5 authentication provides absolutely no protection against the injection of spurious information by an authenticated source.

5.3.2.2 IPsec Authentication

IPsec has been proposed as a stronger authentication (and potentially encryption) mechanism for BGP messages. IPsec Security Associations are applied to individual BGP sessions. However, since a single security mechanism must be mandated in draft standard RFCs, and MD5 TCP authentication is widely implemented and deployed, IPsec is not mandated as an authentication mechanism in the latest draft specification for the BGP protocol. Even IPsec does not provide protection against the injection of spurious information by an authenticated source. Once authenticated, it is assumed that all data from a particular source is valid. The principles of IPsec were described in Chapter 4.

5.3.3 BGP GRACEFUL RESTART

BGP graceful restart is a negotiable capability. During the process of establishing the relationship, two peers indicate to each other that they are each capable of maintaining their forwarding state across a restart of the routing function. It also informs the neighbour that the router is able to send an End-of-RIB marker upon completion of the initial routing table update. This capability can be independently negotiated for each address family.

When the receiving router detects a reset of the TCP session, it starts a restart timer (the value is agreed during initial negotiation) and marks all the routes learned from the restarting routers as stale. Stale routing information should be treated no differently from other routing information. If the restart timer expires, then all stale routes are purged from the routing table of the receiving router.

If the receiving router does not detect the TCP reset, it will receive a new Open message for what it believes to be an existing session. If the two peers have negotiated graceful restart, then the receiving router should reset the original TCP session and use the new session. Since the TCP session has been reset, no Notification message should be sent. The receiving router then goes through the same process of setting the restart timer and marking the routes as stale.

Following the restart, the receiving router will transmit its routes to the restarting router. Once the routing table has been completely sent, the receiving router sends an End-of-RIB marker. The receiving router will also replace any routes marked as stale with identical routes learned from the restarting router.

It is not possible to perform a graceful restart if the TCP session is terminated due to the transmission or receipt of a Notification message. In this case, the standard BGP protocol processes are followed.

While it is clear that graceful restart can significantly reduce the impact of route flapping, it is important to note that it is not without risk. The fact that two routers are maintaining state, which is potentially inconsistent with reality means that it is possible for black holes and transitory routing loops can be created. This risk is generally acceptable given the potentially widespread impact of flapping routes.

It is also worth noting that deploying graceful restart on iBGP sessions in the absence of a similar graceful restart mechanism for the IGP within the autonomous system can reduce the positive benefits of BGP graceful restart.

5.3.4   MULTIPROTOCOL BGP

So far, we've only discussed BGP, as it was originally designed, for IPv4 unicast NLRI. However, BGP4 has now been extended to carry NLRI for IPv4 multicast, and IPv6 unicast and multicast NLRI. It also supports various forms of NLRI associated with VPNs. Rather than trying to cover all of the different types of NLRI in a single location, each is discussed in description of the use of MBGP in the relevant section.

6

Routing Policy

Routing policy is considered by many to be one of the most complex concepts to grasp when designing an IP network. In fact, many engineers find it scary. Implementing policy is not inherently complex or difficult. However, policies are often used to enumerate exceptionally complex ideas and, therefore, it can be exceptionally tricky to write policy that consistently and accurately implements those complex requirements. The way in which you write your policy can dramatically impact the scalability of your network. Well-written policy is easy to understand, easy to update accurately and easy to troubleshoot. Poorly written policy is often none of those things. In general, the simpler the policy, the better. However, it is not always possible to create functional policy that can be implemented simply or elegantly.

This chapter will look at what policy is used for, how it can be implemented and some of the features of the main vendors' syntaxes, which can be utilized to improve the overall scalability of your network.

6.1 WHAT IS POLICY FOR?

The whole point of policy is to be able to influence the way in which your network devices advertise and accept prefixes and thus the way in which they forward traffic. However, this is just the technical element of what policy is for. In the broader sense, policy allows you to implement your business aims.

It is clearly important that you operate your network in such a way to meet the aims of your company's business plan. Otherwise, you are effectively working against the rest of your company! Routing policy provides you with the means to decide which paths are

Designing and Developing Scalable IP Networks   G. Davies
© 2004 Guy Davies   ISBN: 0-470-86739-6

preferred, which prefixes are preferred, and which neighbours are preferred. It also allows you to effectively copy routing information from one routing protocol to another in a controlled fashion. Routing policy can be applied to any routing protocol, although the actual details of what can be done with each different protocol vary. However, the main area in which routing policy is applied, and in which it gets potentially complicated, is in interdomain routing. It is here that the wrong choice among different approaches can cost your company the most money.

6.1.1 WHO PAYS WHOM?

While many engineers, myself included, try to avoid the gory details of how their company's business is operated, the customer–supplier relationships between your network and the networks to which it connects exert a significant influence upon how you implement your policy. It is clearly in your company's interests to pay as little as possible to its suppliers and to obtain as much money as possible from its customers. Routing policy can be a means to improve the ratio of money earned to money expended. This may sound terribly mercenary to some, but any company operating on any other basis will inevitably lose money and eventually fail.

6.2 IMPLEMENTING SCALABLE ROUTING POLICIES

Policy configuration can form a significant proportion of any configuration file in an Internet router. Scalable policies are those that can be easily designed, easily implemented, and easily understood. It is therefore the responsibility of the engineer designing policy to ensure that policies are both easily implemented and easily understood. This latter aspect is often overlooked. However, it is vital that any engineer subsequently looking at the policy does not have to spend hours to understand what the policy configuration is supposed to be doing or, more importantly, what it is *actually* doing.

Modular policy description features allow sections of policy to be created that are extremely generic and can be applied within other policies purely by reference. For example, many policies on your network devices may refer to all addresses within your network. If you create a prefix list, which contains all your internal network addresses, it can be reused in many policy elements. In addition, if you add more addresses to your networks, you merely have to change one prefix list on every routing device rather than changing every relevant policy on every routing device.

Modular policy description features are clearly an extremely desirable feature because they have the potential to help your company to reduce administrative costs. However, depending upon how the syntax is implemented by the vendor and used by the policy designer, there is a risk that it could introduce some complexity. Consider the following two pieces of code (both for JUNOS software but the concept is equally applicable to any other router vendors' configuration languages) (see Table 6.1).

Table 6.1   Comparison of verbose and terse configuration style

Verbose Configuration	Terse Configuration

```
policy-options {                              policy-options {
  prefix-list customer-plus-internal-nets {     prefix-list pl1 {
    10.0.0.0/8;                                   10.0.0.0/8;
    192.168.0.0/16;                               192.168.0.0/16;
  }                                             }
  policy-statement announce-cust-plus-int {     policy-statement ps1 {
    term announce-prefixes {                      term t1 {
      from {                                        from {
        protocol bgp;                                 protocol bgp;
        prefix-list customer-plus-internal-nets;      prefix-list pl1;
      }                                             }
      then accept;                                  then accept;
    }                                             }
    term default-reject {                         term t2 {
      then reject;                                  then reject;
    }                                             }
  }                                             }
}                                             }
```

As you can see, even though the example is relatively simple and both configurations are clearly identical, it is much more obvious that one can understand what this policy does from reading the verbose configuration.

6.3   HOW IS POLICY EVALUATED?

If you do not understand how policy is evaluated, it is impossible to write policy that you can be sure will accurately implement your business aims. Various aspects of policy evaluation are described below.

6.3.1   AND OR OR?

This may seem like a trivial question (or even a nonsensical one if you are not sure where I am heading ...) but in the design and implementation of policy, it is truly fundamental. The question refers to whether particular policy elements are evaluated with a logical AND between them or a logical OR. Unfortunately, the answer to this question is 'it depends'. It depends upon the implementation, and it also depends upon the function within the policy. This may sound like a particularly opaque answer but, hopefully, I will bring a little clarity to the subject below.

6.3.2   THE FLOW OF POLICY EVALUATION

The flow of policy evaluation is an implementation-specific feature. Policy is generally evaluated linearly from start to finish, with the first match causing the associated policy action. Then the policy evaluation is terminated. However, some implementations add the facility to permit subsequent policy elements to be evaluated even after a match has already been found. This can produce exceptionally complex and powerful policy, but with gains in complexity and power comes greater difficulty of understanding. It is, therefore, vital that the implementer of the policy uses whatever means available to ensure the clarity of purpose of his policy.

6.4   POLICY MATCHES

Routing prefixes can be matched on a range of criteria, some of which are specific to individual prefixes (e.g., prefix length, route metric, originating protocol) and some of which are specific to particular routing protocols (e.g., BGP community, IS-IS level). For complete lists of all policy match conditions, refer to the command references of your chosen vendors' software. It is these match criteria that are the basis upon which you can identify the packets or prefixes to which you are going to apply the policy action.

6.5   POLICY ACTIONS

As with policy matches, there is a vast array of policy actions, many of which are specific either to packets or to a particular protocol to which a policy is being applied (e.g. set the BGP community, set the OSPF external type). For complete lists of all policy actions, refer to the command references of your chosen vendors' software.

6.5.1   THE DEFAULT ACTION

The default action is another area in which different vendors have chosen different solutions. Indeed, vendors have made different choices for different elements of their policy tool kits. In all cases, however, the default policy action is either permit or deny. This sounds like an obvious statement, but it is important to realize that the default action is nothing more than one of these two actions. It is essential for the successful implementation of your policy that you know the default action for a given vendor and a given policy element.

6.5.2   ACCEPT/PERMIT, REJECT/DENY, AND DISCARD

Different implementations use different terminology for the same actions. However, the behaviour is fundamentally the same. Accept or permit, if applied to a routing policy, causes prefixes matching the criteria to be installed into the routing table, redistributed to another routing protocol, or announced to a neighbour.

Reject or deny, if applied to a routing policy, causes prefixes matching the criteria not to be installed into the routing table, redistributed to another routing protocol, or announced to a neighbour.

6.6   POLICY ELEMENTS

The functions performed by the combination of access lists, prefix lists, community lists and route maps on Cisco routers are effectively analogous to the functions performed by the combination of firewall filters and policy statements in Juniper Network routers. It is difficult to draw direct comparisons between the specific functions of these features because Juniper Networks and Cisco Systems implementations do not map directly one to the other. There is overlap between different elements of the functionality to give, ultimately, roughly equivalent behaviour.

6.7   AS PATHS

As discussed in Chapter 5, in BGP, every autonomous system (AS) in the path from the origin of a prefix to your network adds its AS number to the AS_PATH attribute of that prefix when it announces it to a neighbouring AS. The AS path length is one of the criteria upon which prefixes are compared when selecting one from a set of otherwise identical prefixes.

In addition to the length of the AS path, it is possible to match specific AS paths using regular expressions and to make policy decisions based on the match. Cisco and Juniper use different syntaxes for their pattern matching. Both syntaxes provide a flexible mechanism for identifying complex patterns in the AS_PATH attribute.

It is also possible to artificially extend the AS path associated with a prefix to make that prefix less attractive than one with an unmodified AS path. This is known as AS path prepending and is achieved by adding your own AS more than once as you announce your prefixes to your neighbours (see Table 6.2).

6.8   PREFIX LISTS AND ROUTE LISTS

Juniper treats prefix lists and route lists slightly differently from how Cisco treats them. Both treat prefix lists differently from route lists. Prefix lists are, as the name suggests, lists of prefixes. In JUNOS software, a prefix list contains just a simple list of prefixes and associated prefix lengths that matches any prefix of equal or longer prefix length to those listed (see Table 6.3).

In IOS software, prefix lists offer slightly more granular control, providing the means to constrain the prefix lengths matched to a particular range (see Table 6.4).

JUNOS route lists, which are analogous to certain functions of Cisco access lists when used in conjunction with distribute lists, may sound similar to prefix lists. However, they

Table 6.2 JUNOS and IOS BGP policy configuration format

Juniper configuration	Cisco configuration
| ```
policy-options {
  policy-statement match-bgp-as-path {
    term reject-prefixes-from-as-path-bad-path {
      from {
        protocol bgp;
        as-path bad-path;
      }
      then reject;
    }
    term default-accept {
      from protocol bgp;
      then accept;
    }
  }
  policy-statement prepend-twice {
    term match-prefix-and-prepend {
      from {
        protocol bgp;
        route-filter 10.0.0.0/8 orlonger;
      }
      then {
        as-prepend "1 1";
        accept;
      }
    }
    term default-accept {
      from protocol bgp;
      then accept;
    }
  }
  as-path bad-path "1000 3456 5678";
}
``` | ```
ip access-list 1 permit 10.0.0.0 255.0.0.0 0.255.255.255 255.0.0.0
!
ip as-path access-list 1 permit ^1000_3456_5678$
!
route-map match-bgp-as-path deny 10
 match as-path 1
!
route-map match-bgp-as-path permit 20
!
route-map prepend-twice permit 10
 match ip address 1
 as-path-prepend 1 1
!
route-map prepend-twice permit 20
``` |

Table 6.3 JUNOS prefix list.

```
policy-options {
  prefix-list mylist {
    permit 192.168.103.0/24;
    deny 0.0.0.0/0;
  }
}
```

Table 6.4 IOS prefix list

```
ip prefix-list 1 permit 192.168.103.0/24 le /30
ip prefix-list 1 deny 0.0.0.0/0
```

are treated very differently and provide different functionality within the overall policy tool kit. In JUNOS route lists, prefixes can be declared as an exact match or declared to match a variety of range schemes. This includes the extra functionality provided by prefix lists in IOS software, but goes way beyond that (see Table 6.5).

IOS access lists (see Table 6.6) provide similar flexibility to JUNOS route lists as well as much of the functionality in JUNOS firewall filters. In fact, access lists and firewall filters are closer in overall functionality than access lists and route lists. Route lists are indeed a subcommand of firewall filters.

Table 6.5   JUNOS firewall filters

```
firewall {
  family inet {
    filter myfilter {
      term permitmyroutes {
        from {
          route-list 192.168.103.0/24 exact accept;
          route-list 192.168.104.0/23 orlonger;
          route-list 192.168.106.0/23 upto 24 reject;
          route-list 192.168.108.0/23 longer;
        }
        then accept;
      }
    }
  }
}
```

Table 6.6   IOS access lists

```
ip access-list permitmyroutes seq 5 permit 192.168.103.0 0.0.0.0 255.255.
   255.0 0.0.0.0
ip access-list permitmyroutes seq 10 permit 192.168.104.0 0.0.1.255 255.255.
   254.0 0.0.1.255
ip access-list permitmyroutes seq 15 deny 192.168.106.0 0.0.1.255 255.255.
   254.0 0.0.1.0
ip access-list permitmyroutes seq 20 permit 192.168.108.0 0.0.1.255 255.255.
   255.0 0.0.1.0
```

Prefix lists and route lists, when combined with BGP routing policy, are very effective tools for the protection of your network against having spurious routing information injected by your neighbours. In an ideal world, all peering sessions would be filtered based on an explicit list of prefixes per neighbour. The administrative load of this kind of filtering can be extremely high, even for customer connections in which you might expect the number of prefixes announced to be quite low. If the filtering were applied to peers, the load would be intolerable unless an automated and totally separate mechanism were available to everyone to announce to the world the set of prefixes they expect to advertise.

6.9   INTERNET ROUTING REGISTRIES

Fortunately for all of us, such a system exists. The Internet Routing Registries (IRRs) provide a globally distributed repository for information that provides the facility to fully describe sets of prefixes and their respective origin ASes being announced by any AS. In order for the repositories to be useful, it is necessary to be able to get the information out of them. Public domain tools exist to retrieve the information from the IRRs and convert it into a set of access lists and route maps for IOS or firewall filters for JUNOS policy statements. This makes it entirely feasible to apply prefix based filtering to all your peers.

However, it is important to note that for this mechanism to be reliable, it is essential that your peers maintain their IRR objects. If they do not keep their IRR objects up to date, you will find that your customers may not be able to reach subsets of their customers. This leads some organizations and individuals to consider that the imposition of this kind of filtering upon your peers is more disruptive that beneficial. There are strongly held opinions on both sides of the argument. Those in favour suggest that if service providers start imposing these constraints on their peers, those peers will 'fall into line' and start keeping their objects up to date. Experience tends to show that this is only partially true. Some service providers just do not keep their objects up to date. Those against the imposition of these constraints point out the woeful state of many IRR objects. Until those objects are complete, there is going to be substantially more hassle than can be justified by the potential benefits.

Note that you should never apply prefix-based filtering to your transit provider. I can guarantee that they will not list the entire Internet in their IRR objects so, if you applied this filtering mechanism to your transit provider, you will not get the transit for which you are paying.

6.10 COMMUNITIES

Communities are an optional attribute associated with one or more prefixes in BGP. They are merely arbitrary tags with no inherent meaning that provide the operator with the ability to group prefixes into 'communities' sharing some abstract characteristic or characteristics. A prefix may be associated with zero or more communities.

The use of communities allows network operators to classify groups of prefixes at one point within the network, usually at the ingress on receipt from a neighbour or at the point at which a prefix is originated, and then take actions based upon the communities elsewhere in the network, usually the egress. So, for example, all routes learned from customers could be tagged with a particular community, known throughout the AS to represent customers. Then, at the egress towards peers, those routes marked as customer routes would be accepted for announcement. However, if a community representing peers throughout the AS were applied to all prefixes learned from peers, those prefixes could be rejected for announcement to other peers on the basis of their community. Table 6.7 shows some example configurations demonstrating the use of communities.

Note that the export policy in the above example towards peers and transit providers is the same (only announce your own customers and your own networks). So, there is no policy named to-transit, the policy named to-peers is used for both.

In addition to user-defined communities, there are some 'well known' communities that can be passed between ASs and that retain the same meaning across AS boundaries. They are as follows:

- *no-export*. Informs the neighbour that this prefix should not be announced beyond the borders of their AS. A prefix with no-export set should never be announced to a neighbour in another AS. However, prefixes with this community are announced to other sub-ASes in a BGP confederation.

Table 6.7 Using BGP communities with JUNOS and IOS

Juniper Configuration	Cisco Configuration

```
protocols {                                  router bgp 1
  bgp {                                       no synchronization
    group all-peers {                         bgp router-id 192.168.0.1
      type external;                          neighbor all-peers peer-group
      import from-peer;                       neighbor all-peers send-community
      export to-peer;                         neighbor all-peers route-map from-peer in
      neighbor 10.0.1.2 {                     neighbor all-peers route-map to-peer out
        peer-as 1000;                         neighbor all-custs peer-group
      }                                       neighbor all-custs send-community
    }                                         neighbor all-custs route-map from-peer in
    group all-custs {                         neighbor all-custs route-map to-peer out
      type external;                          neighbor all-transit peer-group
      import from-customer;                   neighbor all-transit send-community
      export to-customer;                     neighbor all-transit route-map from-transit in
      neighbor 10.255.101.2 {                 neighbor all-transit route-map to-peer out
        peer-as 6666;                         neighbor 10.0.1.2 remote-as 1000
      }                                       neighbor 10.0.1.2 peer-group all-peers
    }                                         neighbor 10.255.101.2 remote-as 6666
    group all-transit {                       neighbor 10.255.101.2 peer-group all-custs
      type external;                          neighbor 172.16.1.2 remote-as 5555
      import from-transit;                    neighbor 172.16.1.2 peer-group all-transit
      export to-peer;                         no auto-summary
      neighbor 172.16.1.2 {                   !
        peer-as 5555;                         ip bgp-community new-format
      }                                       !
    }                                         ip community-list 1 permit 1:2
  }                                           ip community-list 2 permit 1:1
}                                             ip community-list 2 permit 1:2
routing-options {                             ip community-list 2 permit 1:3
  autonomous-system 1;                        !
  router-id 192.168.0.1;                      route-map from-peer permit 10
}                                              set community 1:1
policy-options {                              !
  policy-statement from-peer {                route-map to-peer permit 10
    term t1 {                                  match community 1
      from protocol bgp;                      !
      then {                                  route-map from-customer permit 10
        community set peers;                   set community 1:2
        accept;                               !
      }                                       route-map to-customer permit 10
    }                                          match community 2
  }                                           !
  policy-statement to-peer {                  route-map from-transit permit 10
    term t1 {                                  set community 1:3
      from {
        protocol bgp;
        community customers;
      }
      then accept;
    }
  }
  policy-statement from-customer {
    term t1 {
      from protocol bgp;
      then {
        community set customers;
        accept;
      }
    }
  }
  policy-statement to-customer {
    term t1 {
```

(continued overleaf)

Table 6.7 (*continued*)

Juniper Configuration	Cisco Configuration

```
     from {
        protocol bgp;
        community [ customers peers transit ];
      }
      then accept;
    }
  }
     policy-statement from-transit {
    term t1 {
      from protocol bgp;
      then {
        community set transit;
        accept;
      }
    }
  }
  community peers members 1:1;
  community customers members 1:2;
  community transit members 1:3;
}
```

- *no-advertise*. Informs the neighbour that this prefix should not be announced to any other neighbour, either internal or external.

- *no-export-subconfed*. This is relevant only within BGP confederations. It is used to indicate that this prefix should not be announced beyond the border of any individual member (sub-AS) of the confederation AS.

6.11 MULTI-EXIT DISCRIMINATOR (MED)

The Multi-Exit Discriminator is a metric carried in BGP NLRI and announced to external neighbours. MEDs are only used by routers in a single AS to distinguish between prefixes announced by one other AS.

MED is a metric, which has a NULL default value. This is the source of one problem. Some vendors consider NULL MEDs to be least preferred while others consider NULL MEDs to be most preferred. This contradictory behaviour is problematic when two neighbours use routers from two vendors with differing default behaviour. It is even more problematic if a single AS uses routers from two vendors with differing default behaviour.

MEDs can have a value between 0 and $2^{32}-1$.

6.12 LOCAL PREFERENCE

Local Preference (LOCAL_PREF) is a locally defined value that represents the preference of a prefix within a single AS. LOCAL_PREF allows the administrator of a network to prefer one source of NLRI over another on a per prefix basis (although the configuration of LOCAL_PREF values on a per prefix basis would be extremely complex and tedious).

Unlike most other metric like values, a high LOCAL_PREF is more preferred than a low one. The value of LOCAL_PREF can be anywhere between 0 and 2^{32}-1. The default value is 100.

6.13 DAMPING

Damping is the mechanism whereby prefixes that have been repeatedly announced and withdrawn over a limited period are marked unusable until the prefix has been present continuously for a period of time.

The mechanism is based on a figure of merit associated with each prefix. Each flap (announcement and withdrawal) of the prefix causes the figure of merit to be increased by a penalty. If the figure of merit rises above a threshold, the prefix is declared ineligible for use. The figure of merit is decayed exponentially, based on a half life, while the prefix remains stable. The prefix is marked eligible for use again only when the figure of merit has fallen below another threshold.

The aim of damping is to contain the flapping prefixes, which consume resources in Internet routers and, particularly with older designs, significantly impair forwarding performance.

Experience has shown that generally longer prefixes (less aggregated prefixes) are less stable than shorter prefixes (highly aggregated prefixes). Therefore, in order to gain the greatest impact, service providers often damp longer prefixes much more aggressively than shorter prefixes. Thus, in order to avoid having your prefixes damped, it is advisable to aggregate your prefixes as much as possible. This not only makes it less likely that your prefixes will be damped, but also reduces the total number of prefixes in the global routing table. However, it is not realistic to expect to maintain absolute aggregation. For example, any customer with addresses from your provider aggregatable address space that subsequently dual-homes to a second ISP will have to have their prefix deaggregated.

It is common practice to generate static aggregate routes that are redistributed into BGP. It is these aggregates that are announced to your neighbours. These routes are extremely stable, normally only disappearing if the router on which they are configured actually crashes. By generating the aggregate on multiple boxes in your network, you overcome this problem.

It has been noted that the use of inconsistent default parameters for damping by the two main vendors of Internet core routers has led in some cases to counter-productive results (see Table 6.8). Indeed, it has been observed that oscillations can propagate longer than they might otherwise have done in an undamped environment.

This issue was raised by Christian Panigl at two consecutive RIPE meetings and was the stimulus for the routing working group of RIPE to produce a document (RIPE-229[1]) making recommendations for the parameters that should be applied to route flap damping. The recommended values differ quite significantly from the default values on the various vendors' routers, which also differ significantly from each other. Before implementing route flap damping, it is well worth reading RIPE-229, then read it again.

---

[1] RIPE-229: *RIPE Routing-WG Recommendations for Coordinated Route-Flap Damping Parameters* Christian Panigl, Joachim Schmitz, Philip Smith, Cristina Vistoli

Table 6.8   Configuring BGP route flap damping with JUNOS and IOS

Juniper configuration	Cisco configuration
| ```
protocols {
  bgp {
    import [ damp-lite damp-norm
      damp-hard ];
    damping;
  }
}
policy-options {
  damping lite {
    half-life 5;
    max-suppress 30;
    reuse 1000;
    suppress 4000;
  }
  damping norm {
    half-life 15;
    max-suppress 60;
    reuse 750;
    suppress 3000;
  }
  damping hard {
    half-life 30;
    max-suppress 120;
    reuse 500;
    suppress 2000;
  }
  policy-statement damp-lite {
    term t1 {
      from {
        protocol bgp;
        route-filter 1.0.0.0/8
          orlonger;
      }
      then {
        damping lite;
        accept;
      }
    }
  }
  policy-statement damp-norm {
    term t1 {
      from {
        protocol bgp;
        route-filter 2.0.0.0/8
          orlonger;
      }
      then {
        damping norm;
``` | ```
router bgp 1
 bgp dampening 5 30 1000 4000
    route-map damp-lite
 bgp dampening 15 60 750 3000
    route-map damp-norm
 bgp dampening 30 120 500 2000
    route-map damp-hard
 !
ip prefix-list 1 permit 1.0.0.0/8
    le 32
ip prefix-list 2 permit 2.0.0.0/8
    le 32
ip prefix-list 3 permit 3.0.0.0/8
    le 32
 !
route-map lite permit 10
 match ip prefix 1
 !
route-map norm permit 10
 match ip prefix 2
 !
route-map hard permit 10
 match ip prefix 3
``` |

(continued overleaf)

Table 6.8 (*continued*)

Juniper configuration	Cisco configuration

```
      accept;
    }
  }
}
policy-statement damp-hard {
  term t1 {
    from {
      protocol bgp;
      route-filter 3.0.0.0/8
        orlonger;
    }
    then {
      damping hard;
      accept;
    }
  }
}
}
```

6.14 UNICAST REVERSE PATH FORWARDING

Unicast reverse path forwarding (RPF) is a policy mechanism that is usually used to provide a degree of protection against spoofing of source addresses (a common feature of DoS packets). It works by checking the source of the inbound packet. If the route back to the source of the packet does not travel directly via the interface over which the packet was received, the packet is dropped.

While this is an attractive mechanism, particularly in today's Internet where DoS attacks are a fact of daily life, it is essential that you take care when implementing it. If a network has only one path to the Internet and that path is via you, then implementing unicast RPF back towards that customer is safe. However, if there is more than one path via which a network can be reached, there is a risk of asymmetric routing. For example, suppose a multihomed customer is connected to you and to another ISP with which you peer. Your service is more expensive than the other ISP so your customer prefers to send traffic via the other ISP to reduce their costs. However, your policy is to prefer routes learned from customers over routes learned from peers. Therefore, when you receive a packet from that customer via your peer, if you have applied unicast RPF towards the peer, you will drop the packet because the path back to the customer is across the directly connected link.

For this reason, it is generally a bad idea to apply unicast RPF to anyone other than your singly homed customers. It is almost impossible to guarantee symmetric routing between your network and any other multihomed network since other networks will apply their own policy to meet their own business needs. Those business needs may be directly contradictory to your own.

Applying different configurations to different classes of customers is also potentially problematic. Keeping configurations as generic as possible reduces the problems associated with the application of the 'wrong' configuration to a particular customer.

In order to overcome some of the constraints of strict unicast RPF while relinquishing some of the absolute benefits, loose unicast RPF was proposed. Instead of checking whether the route back to the source is via the interface over which the packet was received, loose unicast RPF simply checks that a route back to the source exists in the forwarding table, irrespective of the interface via which the source is reached. The main benefit of this has already been described above. However, the main disadvantage, relative to strict unicast RPF, is that if packets are sent with their source address spoofed to a genuine address present on the Internet, then loose unicast RPF will not detect them. Also, in any environment in which there is a default route, loose unicast RPF will always accept a packet because it will always have a path, via the default route, to the source. Some people consider that these problems significantly reduce the benefit of loose unicast RPF almost to the point of being useless.

6.15 POLICY ROUTING/FILTER-BASED FORWARDING

Policy routing or filter-based forwarding is the forwarding of traffic based on policies or filters, rather than using the default routing table to decide the forwarding path.

Policy routing can be used to overcome many issues and provide functionality otherwise not possible with the standard routing protocols and processes. An example of this is the provisioning of self-selection transit exchange services. Consider a 'transit exchange' with connections to several large ISPs (the upstream providers), each of whom offers full transit. In addition, many customers connect to the exchange and through some unspecified mechanism, select their preferred upstream provider based on some arbitrary criteria (e.g. cost per Megabyte, latency across the upstream provider's network, packet loss across the upstream provider's network). Assuming that each upstream provider uses BGP to announce routes to the exchange, default routing processes would select one best path from among all the alternatives provided by each upstream provider and would announce only that route. Using policy routing, the exchange could force traffic to be sent to the selected upstream provider based on the customer from which the traffic was sent, regardless of whether that provider has the best path to the destination. Reverse traffic could be forced back through the correct upstream provider by applying filters to the outbound BGP announcements so that routes from only those customers using the upstream provider for outbound traffic are sent to each upstream provider.

6.16 POLICY RECOMMENDATIONS

6.16.1 POLICY RECOMMENDATIONS FOR CUSTOMER CONNECTIONS

If you wish to earn more from your customers, you should ensure that they receive as much traffic from you as you can validly send to them and that you make it as attractive as possible to them to send traffic your way. This is particularly true for those customers who pay for their service based on traffic volume. Therefore, a prefix announced by a customer should be preferred over the same prefix from any other source (see Table 6.9). That can easily be achieved by setting a high LOCAL_PREF on all routes learned from the customer. Since the default value of LOCAL_PREF is 100 and the maximum value is $2^{32}-1$, there is plenty of scope for applying different values to different customers. Generally, you announce all routes in your routing table to your customers so it is unnecessary to apply any policy to select those routes.

6.16.2   POLICY RECOMMENDATIONS FOR PEERING CONNECTIONS

Peerings are generally a zero cost exchange of routing information between two networks over which the neighbours announce their own networks and those of their customers (but not those of other peers and transit providers). Because traffic to or from these neighbours neither earns nor costs money, prefixes learned from these neighbours should be preferred over those learned from transit providers but should be less favoured than otherwise identical prefixes learned from customers.

Normally, peers only receive prefixes originated locally or learned from your customers. In order to implement this policy, it is necessary to be able identify routes originated locally or learned from customers, peers and transit providers (see below). This can be achieved with communities. By marking each prefix on ingress (or at the point of origin) with a community representing the source of the prefix, it is possible to use that community at the point of egress to select the routes to be announced based on the source of the prefix (see Table 6.10).

6.16.3   POLICY RECOMMENDATIONS FOR TRANSIT CONNECTIONS

The relationship with a transit provider is invariably one for which you will pay, often a large amount of money and often on the basis of the volume of traffic they transport on your behalf. Therefore, it is important to minimize the volume of traffic sent to your transit providers (without reducing the overall connectivity of your network). This means that prefixes learned from transit providers should be the least preferred of a set of otherwise identical prefixes learned from a customer, a peer, and a transit provider.

As with peers, transit providers normally only accept prefixes originated locally and from your customers (see Table 6.11). In order to minimize the use of routes learned from transit providers, it is simply necessary to set the LOCAL_PREF on routes learned from that neighbour to a low value. Since the default LOCAL_PREF is 100, it is simply necessary to set LOCAL_PREF to a value of less than 100.

Table 6.9 Limited sample BGP configuration to customers using JUNOS and IOS

Juniper Configuration	Cisco Configuration

```
interfaces {                          ip cef distributed
  e1-0/0/0 {                          !
    unit 0 {                          interface serial 0/0/0
      family inet {                    ip address 10.255.101.1
        address 10.255.101.1/30;         255.255.255.252
        rpf-check fail-filter           ip verify unicast reverse-path
          cust-rpf;                     !
      }                               ip prefix-list customerA permit
    }                                    192.168.0.0/24
  }                                   ip prefix-list customerA permit
}                                        192.168.2.0/23
routing-options {                     ip prefix-list customerA2 permit
  aggregate route 158.43.0.0/16;         172.16.0.0/16
  aggregate route 146.188.0.0/16;     ip prefix-list local-aggregates
}                                        permit 172.20.0.0/15
firewall {                            !
  filter cust-rpf {                   ip community-list 1 permit 1:1
    term customer-networks {          ip community-list 2 permit 1:2
      from {                          ip community-list 3 permit 1:3
        source-address {              ip community-list 199 permit 1:1000
          192.168.103.0/24;           !
        }                             ! accept known prefixes and
      }                                  aggregate
      then accept;                    route-map from-customers-template
    }                                    permit 10
    term block-the-rest {              match ip prefix customerA
      then {                           set community 1:2
        count rpf-rejects;             set local-preference 120
        reject;                       !
      }                               ! accept known prefixes and leak
    }                                 route-map from-customers-template
  }                                      permit 15
}                                      match ip prefix customerA2
policy-options {                       match community-list 199
  prefix-list local-aggregates {       set community 1:2 additive
    158.43.0.0/16;                     set local-preference 120
    146.188.0.0/16;                   !
  }                                   ! default reject
  policy-statement                    route-map from-customers-template
    from-customers-template {            deny 100
    term accept-known-prefixes-and-   !
      aggregate {                     ! accept peer routes
      from {                          route-map to-customers permit 10
        protocol bgp;                  match community-list 1
        route-filter 192.168.0.0/24   !
          exact;                      ! accept customer routes
```

(continued overleaf)

Table 6.9   (*continued*)

Juniper Configuration	Cisco Configuration

```
    route-filter 192.168.2.0/23        route-map to-customers permit 15
      exact;                            match community-list 2
  }                                    !
  then {                              ! accept transit routes
    community set customers;          route-map to-customers permit 20
    local-preference 120;              match community-list 3
    accept;                           !
  }                                   ! accept local aggregate prefixes
}                                     route-map to-customers permit 25
term accept-known-prefixes-and-        match ip prefix local-aggregates
    leak {                            !
  from {                              ! default reject
    protocol bgp;                     route-map to-customers deny 100
    route-filter 172.16.0.0/16
        exact;
    community please-leak-this;
  }
  then {
    community add customers;
    local-preference 120;
    accept;
  }
}
  term default-reject {
    then reject;
  }
}
policy-statement to-customers {
  term accept-peer-routes {
    from {
      protocol bgp;
      community peers;
    }
    then accept;
  }
  term accept-customer-routes {
    from {
      protocol bgp;
      community customers;
    }
    then accept;
  }
  term accept-transit-routes {
    from {
      protocol bgp;
      community transit;
    }
```

(*continued overleaf*)

Table 6.9 (*continued*)

Juniper Configuration	Cisco Configuration

<pre>   then accept;
  }
  term accept-local-aggregate-
    prefixes {
    from {
      protocol aggregate;
      prefix-list
        local-aggregates;
    }
    then accept;
  }
  term default-reject {
    then reject;
  }
}
community peers members 1:1;
community customers members 1:2;
community transit members 1:3;
community please-leak-this
   members 1:1000;
}</pre>	

Table 6.10 Limited sample BGP configurations to peers using JUNOS and IOS

Juniper Configuration	Cisco Configuration

<pre>Routing-options {
  aggregate route 158.43.0.0/16;
  aggregate route 146.188.0.0/16;
}
policy-options {
  prefix-list local-aggregates {
    158.43.0.0/16;
    146.188.0.0/16;
  }
  policy-statement
    from-peers-template {
    term accept-known-prefixes-and-
      as-paths {
      from {
        protocol bgp;
        route-filter
          192.168.100.0/24 exact;
        route-filter
          192.168.200.0/23 exact;
        as-path this-peer;
      }</pre>
	<pre>ip prefix-list peerA permit
	   192.168.100.0/24
	ip prefix-list peerA permit
	   192.168.200.0/23
	!
	ip as-path access-list 1 permit
	   ^1000_
	!
	ip community-list 1 permit 1:1
	ip community-list 2 permit 1:2
	ip community-list 3 permit 1:3
	!
	! accept known prefixes and AS
	   paths
	route-map from-peers-template
	   permit 10
	 match ip prefix peerA
	 match as-path 1
	 set community 1:1
	 set local-preference 105
	!</pre>

(*continued overleaf*)

Table 6.10   (*continued*)

Juniper Configuration	Cisco Configuration

```
    then {                          ! default reject
      community set peers;          route-map from-peers-template deny
      accept;                         100
    }                               !
  }                                 ! accept customer routes
  term default-reject {             route-map to-peers permit 10
    then reject;                     match community-list 2
  }                                 !
}                                   ! need mechanism to match
policy-statement to-peers {            prefix-list
  term reject-peer-routes {         route-map to-peers permit 15
    from {                           match ip prefix local-aggregates
      protocol bgp;                 !
      community peers;              ! default reject
    }                               route-map to-peers deny 100
    then reject;
  }
  term reject-transit-routes {
    from {
      protocol bgp;
      community transit;
    }
    then reject;
  }
  term accept-customer-routes {
    from {
      protocol bgp;
      community customers;
    }
    then accept;
  }
  term accept-local-aggregate-
     prefixes {
    from {
      protocol aggregate;
      prefix-list
        local-aggregates;
    }
    then accept;
  }
  term default-reject {
    then reject;
  }
}
as-path this-peer "1000 .*";
community peers members 1:1;
community customers members 1:2;
community transit members 1:3;
}
```

Table 6.11 Limited sample BGP configurations to transit providers using JUNOS and IOS

Juniper Configuration	Cisco Configuration

```
routing-options {                         ip prefix-list local-agg permit
  aggregate route 158.43.0.0/16;             158.43.0.0/16
  aggregate route 146.188.0.0/16;         ip prefix-list local-agg permit
}                                            146.188.0.0/16
policy-options {                          !
  prefix-list local-aggregates {          ip community-list 1 permit 1:1
    158.43.0.0/16;                        ip community-list 2 permit 1:2
    146.188.0.0/16;                       ip community-list 3 permit 1:3
  }                                       !
  policy-statement                        ! accept everything from transit
    from-transit-template {               route-map from-peers-template
    term accept-everything-from-            permit 10
      transit {                            set community 1:3
      from protocol bgp;                  !local-preference defaults to 100
      then {                              !
        community set transit;            ! default reject
        accept;                           route-map from-peers-template deny
      }                                       100
    }                                     !
    term default-reject {                 ! accept customer routes
      then reject;                        route-map to-peers permit 10
    }                                      match community-list 2
  }                                       !
  policy-statement to-peers {             ! need mechanism to match
    term reject-peer-routes {                prefix-list
      from {                              route-map to-peers permit 15
        protocol bgp;                      match ip prefix local-aggregates
        community peers;                  !
      }                                   ! default reject
      then reject;                        route-map to-peers deny 100
    }
    term reject-transit-routes {
      from {
        protocol bgp;
        community transit;
      }
      then reject;
    }
    term accept-customer-routes {
      from {
        protocol bgp;
        community customers;
      }
      then accept;
    }
    term accept-local-aggregate-
      prefixes {
```

(continued overleaf)

Table 6.11   (*continued*)

Juniper Configuration	Cisco Configuration

```
    from {
      protocol aggregate;
      prefix-list
        local-aggregates;
    }
    then accept;
  }
  term default-reject {
    then reject;
  }
}
community peers members 1:1;
community customers members 1:2;
community transit members 1:3;
}
```

If you have multiple connections to a single transit provider, you may find that it is desirable to prefer one link over another. Again, the routes learned over the preferred link can be given a higher LOCAL_PREF (still less than 100). In order to ensure that the preferred link is used by your transit provider, it may be necessary to apply Multi-Exit Discriminators (MED) to prefixes you announce to them. MEDs are used to influence the routing preferences of your peers. They are effectively a metric carried with each prefix announced to your neighbours. In exactly the same way as metrics work in other routing protocols, the lower the MED, the more preferred the path. Therefore, you will need to set the MED to a low value on the prefixes announced over the more preferred link.

In some cases, your transit providers may provide a set of publicly documented communities that you may set in order to influence their routing policy in a more granular way. If such communities are provided, this is usually a more functional mechanism for influencing the flow of traffic from your neighbour into your network.

6.17   SIDE EFFECTS OF POLICY

It is important to realize that while the operator of any AS writes policies for the purpose of implementing their own business aims, that policy can have unexpected and, occasionally undesirable, side effects beyond the boundaries of the AS in which it is implemented. A particular example of this is illustrated in RFC 3345.[2] In this case, differences in the way BGP policy is implemented by adjacent service providers and the partial visibility of egress points caused by use of route reflectors or BGP confederations can lead to

---

[2] RFC 3345: 'Border Gateway Protocol (BGP) Persistent Oscillation Condition', McPherson, D., Gill, V., Walton, D. and Retana, A., August 2002.

a condition in which a prefix never finds a stable preferred path. This means that the routers involved in BGP are constantly flapping, updating their routing and forwarding information with respect to their neighbour's prefixes, which can be a significant drain on the resources not only of the directly involved routers, but also of those that receive the affected information from the directly affected routers.

In Figure 6.1, taken from the example of Type I churn with route reflectors in RFC 3345, AS 1 contains two route reflection clusters, each with a single reflector (identified in Figure 6.1 as core routers). The numbers on the links within AS 1 represent the IGP metric on the link. The numbers on the links between AS 1 and the other ASes represent the MED associated with the prefix 10.0.0.0/8 as announced to AS 1.

If we start with the RR in cluster 1 with the following routing information, as shown in Table 6.12. In the initial state, the RR prefers the route via AS 10. The RR recognizes that this is incorrect based on the IGP metric. The RR selects the route via AS 6, and announces that prefix to the RR for cluster 2. The RR in cluster 2 has the following prefixes in its routing table (see Table 6.13).

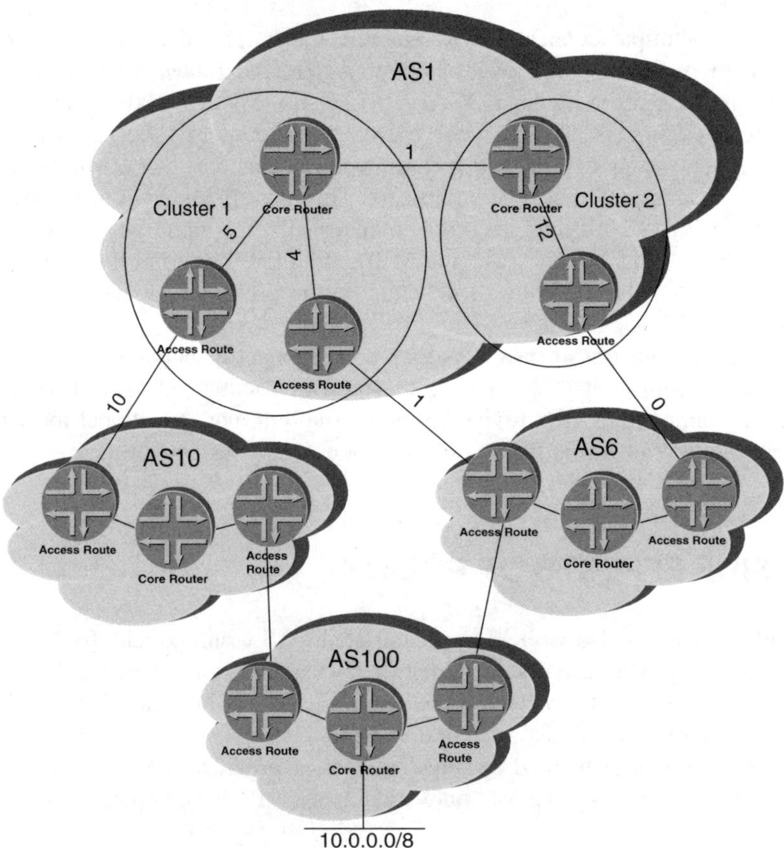

Figure 6.1 Example network for persistent oscillation

Table 6.12   Initial routing state in cluster 1

AS Path	MED	IGP metric to BGP next hop
6 100	1	5
10 100	10	4

Table 6.13   Initial routing state in cluster 2

AS Path	MED	IGP Cost to BGP next hop
6 100	0	12
6 100	1	5

Table 6.14   Modified routing state in cluster 1

AS Path	MED	IGP metric to BGP next hop
6 100	0	13
6 100	1	4
10 100	10	5

Table 6.15   Modified routing state in cluster 2

AS Path	MED	IGP metric to BGP next hop
6 100	0	12
10 100	10	5

The RR for cluster 2 prefers the route via the local client to AS 6, based on the lower MED. It sends a withdrawal/update to the RR for cluster 1 indicating the new preferred path via AS 6. The RR for cluster 1 now has the following prefixes in its routing table (see Table 6.14).

The first prefix via AS 6 is preferred based on the MED. It is then compared with the path via AS 10. The latter wins based on the lower IGP metric to the BGP next hop. The RR for cluster 1 sends a withdrawal/update to the RR for cluster 2 with the new preferred prefix. This means that the RR for cluster 2 has the following prefixes in its routing table (see Table 6.15).

The RR for cluster 2 now selects the route via AS 10 based on the lower IGP metric. It sends a withdrawal/update back to the RR for cluster 1, which now has the prefixes shown in Table 6.12. And we are back to the start of the process. This process will not stabilize without external intervention. The following are possible mechanisms that might help in the prevention of this oscillation:

- Ensure that the intercluster links have significantly higher metrics than the intracluster links. This is by far the cleanest workaround, because it is entirely within the control of the affected network operator and effectively prevents the oscillation.

- Do not accept MEDs from peers. This is within the control of the affected network operator. However, it defeats the object of MEDs.

- Apply other attributes, higher up the decision process, to inbound prefixes. See Table 6.16, for the order in which Cisco Systems and Juniper Networks routers evaluate the various attributes. This mechanism achieves the same ends as the previous option and imposes a much greater administrative load. Therefore, if you are considering this option, just use the previous option.

- Always compare MEDs, regardless of the AS from which the prefix was learned. Given the variety of mechanisms used by service providers to derive MEDs, this is probably ill advised. It also puts even greater control of the flow of traffic out of your network and into the hands of your neighbours. This is also contrary to the purpose of MEDs, which were explicitly devised to provide differentiation between multiple paths learned from a single neighbour AS.

- Use a full internal BGP mesh. This overcomes the problem entirely. However, this is unlikely to be a realistic solution in any but the smallest, most static networks.

The problem described above is also the same if an AS is broken up using confederations. For a full description of this and another variation on this problem, see RFC 3345. Table 6.16 compares the active route selection processes in JUNOS and IOS.

So, as you can see, you have to be aware of how policy decisions made by others might affect the decisions you have to make.

Table 6.16   Comparison of the active route selection processes in JUNOS and IOS

Juniper	Cisco
1. Prefer the prefix with the lowest preference value.	1. Prefer the prefix with the lowest administrative distance.
2. If both prefixes have the same preference value (i.e. they are being learned from the same routing protocol), prefer the prefix with the highest LOCAL_PREF. For non-BGP paths, prefer the prefix with the lowest preference 2 value.	2. Prefer the prefix with the highest LOCAL_PREF.
3. Prefer the prefix with the shorter AS PATH.	3. Prefer the prefix with the shorter AS PATH.
4. Prefer the prefix with the lower origin code. IGP < EGP < incomplete.	4. Prefer the prefix with the lower origin code. IGP < EGP < incomplete.
5. For prefixes learned from the same external BGP neighbour AS, prefer the prefix with the lowest MED is preferred. The requirement for both prefixes to have been learned from the same neighbour AS can be overridden with the BGP command `path-selection always-compare-med`.	5. For prefixes learned from the same BGP neighbour, prefer the prefix with the lowest MED.
6. Prefer strictly internal routes (i.e. routes learned via an IGP).	6. Prefer the strictly internal path (i.e. those learned from an IGP).
7. Prefer strictly external routes (i.e. routes learned directly from an external BGP neighbour rather than an internal BGP neighbour).	7. Prefer strictly external paths (i.e. those learned from an external BGP neighbour rather than an internal BGP neighbour).
8. Prefer the prefix with the lowest IGP metric to the BGP next hop.	8. Prefer the prefix with the shortest IGP metric to the BGP next hop.
9. Prefer the prefix with the shortest route reflection cluster list. If no cluster list is present, this is considered to have a length of zero.	9. Prefer the prefix with the shortest route reflection cluster list.
10. Prefer the prefix learned from the neighbour with the lowest BGP router ID.	10. Prefer the prefix learned from the neighbour with the lowest BGP router ID.
11. Prefer the prefix learned from the neighbour with the lowest peer IP address.	11. Prefer the prefix learned from the neighbour with the lowest peer IP address.

7

Multiprotocol Label Switching (MPLS)

Back in the mid to late 1990s, when most routers were predominantly based on software forwarding rather than hardware forwarding, a number of vendors devised proprietary mechanisms to switch packets far more efficiently than was possible with forwarding based entirely on hop-by-hop longest match IP address lookups. Various aspects of these proprietary mechanisms were effectively merged and developed by the MPLS working group at the IETF and produced what we know today as MPLS.

Subsequent developments in hardware-based router design have made these forwarding performance improvements unnecessary, since these hardware-based routers are capable of performing longest-match IP lookups at exactly the same speed as fixed-length label lookups.

MPLS' forwarding decision is based entirely on the contents of a fixed-length label assigned to the packet at the ingress of the network. This label forms a part of the 32-bit MPLS header as shown in Figure 7.1.

The label is 20-bits long giving 1048576 label combinations. There are three experimental (EXP) bits for which only one purpose has currently been proposed (class of service marking). The S bit indicates whether this label is at the bottom of the label stack (1 indicates that this label is the bottom of the stack, 0 indicates it is not). Finally, there is an 8-bit TTL that is used to limit the impact of any routing loops.

The fact that a label is defined for each hop effectively means that the entire path is predefined creating a unidirectional 'circuit' from ingress to egress. These 'circuits' are called Label Switched Paths (LSPs). This circuit-like behaviour makes MPLS ideal for a number of functions including integration of IP networks with ATM networks,

Designing and Developing Scalable IP Networks   G. Davies
© 2004 Guy Davies   ISBN: 0-470-86739-6

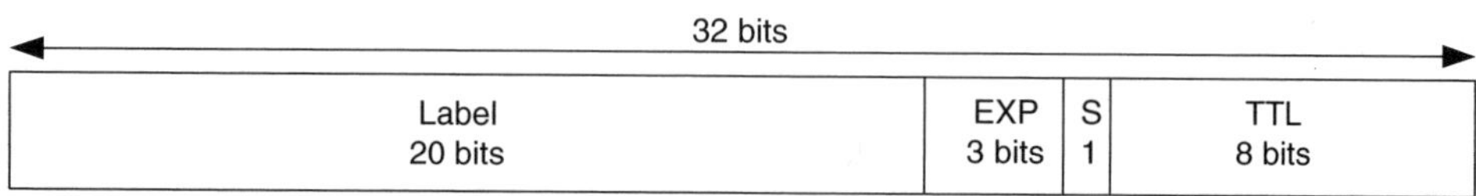

Figure 7.1 MPLS label

traffic engineering of IP networks and the provision of Layer 2 and Layer 3 Virtual Private Networks.

Any device which supports MPLS is known as a Label Switching Router (LSR). These devices can include both traditional routers, ATM switches and, with the introduction of Generalized MPLS (described towards the end of this chapter), Optical Cross Connects and Add-Drop Multiplexers. This is the basis of the integration of frame-based IP networks with ATM and optical transmission networks.

7.1 TRAFFIC ENGINEERING

LSPs have many of the characteristics of a circuit and can, to some extent, be manipulated in much the same way as an ATM or Frame Relay PVC. This behaviour can be exploited for IP traffic engineering.

Traffic engineering is all about the ability to place traffic onto particular paths so that no resource in the network is overloaded while other resources are under-utilized. As stated earlier in this book, it is essential in today's market to be able to make the most of your assets. If you can spread traffic through your network across all available paths, then you should need to buy new capacity later than would be necessary if traditional routing were used. This is a significant benefit to your company's cash flow.

In addition to the benefits associated with placing traffic on particular links for the purposes of load sharing, it can also be useful to be able to place traffic associated with different classes of service onto different links. For example, consider two locations on your network between which there is a large aggregate amount of data, of which a small proportion is associated with a latency intolerant application. Your customers are prepared to pay more for the latency intolerant traffic to be given a higher priority as long as the latency intolerant packets get through within the application's requirements. You have a very expensive link directly connecting the ingress and egress points on your network but it has limited capacity. An alternative path goes via several cities and is significantly longer than the direct path. However, it has very much larger capacity and is much cheaper.

You can create two LSPs between the ingress and egress, one of which is forced via the expensive, low latency link, and the other, which is forced via the low cost, higher latency link (see Figure 7.2). Traffic for the low latency application is placed on the direct path. All other traffic is placed on the cheaper path.

Traffic engineering requires the distribution of extra information about links over and above the basic metric information carried by traditional IP routing protocols. This information includes available bandwidth, reserved bandwidth, over-subscription rates and

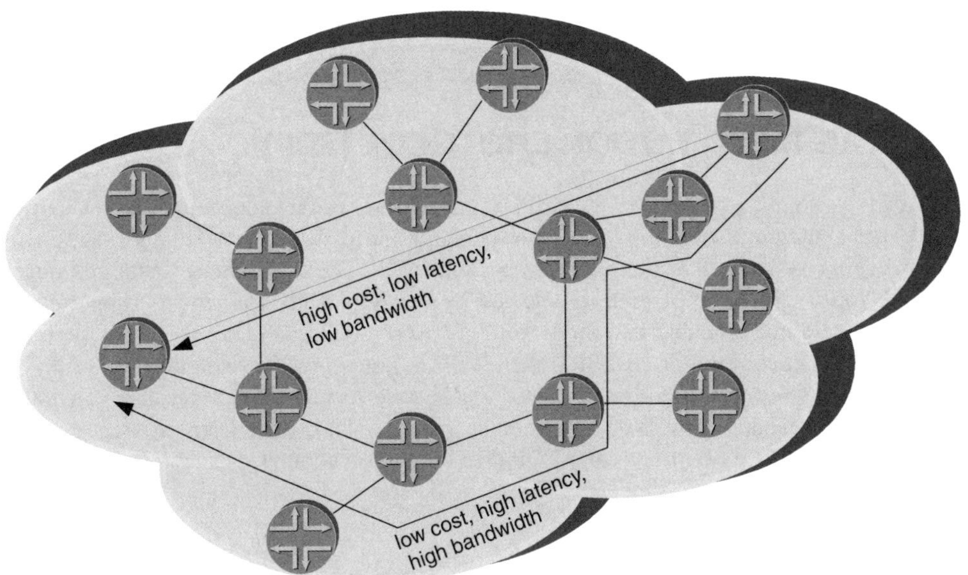

Figure 7.2 Traffic engineering with MPLS

administrative colours. It is this information, in combination with the standard metrics associated with routing protocols, which is used to select the 'best' path. The process of selecting the best path using this information is known as constrained shortest path first (CSPF). As the name suggests, it is based on the same shortest path first (SPF) algorithm used by OSPF and IS-IS with the ability to consider added constraints (bandwidth, administrative colour, etc.) in the calculation of the shortest path. Put simply, links that do not comply with the required constraints are pruned from the tree of available links prior to the calculation of the shortest path based on the link metrics. In fact, CSPF is based on extended versions of OSPF or IS-IS. These are described below in more detail.

7.2 LABEL DISTRIBUTION PROTOCOLS

While it is possible to statically define labels for all LSPs on the LSRs throughout the network, this is the equivalent of operating a network entirely using static routes. It works but it is not flexible, not resilient and most certainly not scalable.

Label distribution protocols (note the lower case) are mechanisms whereby one LSR informs another LSR of the label to be associated with a particular forwarding equivalence class (FEC). This is a fundamental part of the MPLS system. MPLS can be operated without a label distribution protocol by the configuration of static label pushes, swaps and pops. However, this is certainly not a scalable way of doing things and, other than as a means of familiarizing yourself with the label swapping process in the lab, static label mappings are of little practical value. So, you are left with a requirement for a label distribution protocol.

7.3   TAG DISTRIBUTION PROTOCOL (TDP)

One of the original, proprietary pre-MPLS protocols was developed by Cisco and was called tag switching. TDP is Cisco's protocol for the distribution of 'tags' used in its tag switching system. It works on a hop-by-hop basis. The tag switched paths produced by TDP follow the shortest path as selected by the IGP. Tag switching is supported on Cisco's routers in frame and cell mode and on Cisco's ATM switches in cell mode only.

Unless you have only Cisco LSRs, then TDP is not a practical solution. Even if you do have only Cisco LSRs today, it is not really wise to use TDP because it constrains you if you subsequently wish to use another vendor's hardware. Cisco routers can run TDP on one interface of a router and LDP (see below) on another, providing a conversion interface between the two protocols.

7.4   LABEL DISTRIBUTION PROTOCOL (LDP)

LDP is a protocol similar to, but incompatible with, TDP. It also works on a hop-by-hop basis producing LSPs, which follow the shortest path from ingress to egress as selected by the IGP. It was designed by the MPLS working group of the IETF. LDP is widely supported by router vendors. It is also supported on ATM LSRs providing the means for integrating signalling of LSPs from end to end across both frame and cell-based networks.

If you have no traffic engineering requirements, it is generally simpler to implement LDP than RSVP-TE (see below). An LSR running LDP discovers its neighbours and automatically distributes a label for each address for which it has received a label. Cisco routers by default also create a label for every attached interface. Juniper routers only produce a label for the loopback interfaces. Since labels are not created for the links to your eBGP neighbours, you must set next-hop-self on iBGP sessions or the next hop will not be accessible. If you have a combination of Cisco and Juniper routers as PE, then it is advisable to decide which model you will use and ensure that you use it consistently. Under these circumstances, my advice would be to limit the Cisco PE to advertising only the loopback address (see Table 7.1).

Table 7.1   IOS configuration to constrain LDP advertisements to loopback0 only

```
mpls ip
mpls ldp advertise-labels for 1
!
interface loopback0
 ip address 192.168.255.1 255.255.255.255
 mpls ip
!
access-list 1 permit 192.168.255.1 0.0.0.0
access-list 1 deny any
```

7.4.1   LDP GRACEFUL RESTART

As with several of the routing protocols described in Chapter 5, it is desirable to be able to support graceful restart for LDP sessions. Graceful Restart for LDP is described in RFC 3478.

If the LDP control plane in an LSR restarts, it is preferable not to have to tear down all the LSPs traversing that LSR, recalculate a shortest path and resignal the new LSPs avoiding the restarting LSR then, upon recovery of the LSR, be left with non-optimal LSPs until an optimization of the LSPs is carried out. Instead, in cases where the forwarding plane can be maintained across the temporary failure of the control plane, it is better simply to continue forwarding traffic using exactly the same mappings of incoming label, incoming interface, outgoing label, outgoing interface/next-hop.

Adjacent LSRs signal their ability to maintain their MPLS forwarding state derived from LDP by setting the fault tolerant (FT) TLV during the initial exchanges of the LDP protocol. This FT TLV contains two items, a Reconnect Timeout and a Recovery Time.

The Reconnect Timeout is the time in milliseconds within which the restarting router can restart its control plane and establish an LDP session to the receiving router such that LDP information can be exchanged. The receiving router should maintain the MPLS forwarding state for the period specified in the Reconnect Timeout. If the Reconnect Timeout specified by the restarting router is zero, it indicates that the restarting router cannot maintain its MPLS forwarding state across the control plane restart. This sounds rather strange, but it indicates that this router is capable of responding correctly if a neighbour can maintain its forwarding state across the restart.

If the restarting router can maintain its forwarding state, it sets a configurable timer and marks all its forwarding entries as stale. As it receives updates following the restart, the restarting router replaces stale entries with updated entries. When this timer expires, any remaining stale entries are purged.

Similarly, when the receiving router notices that the LDP session with its neighbour has been lost, it checks whether that neighbour is capable of maintaining its state across the restart. If so, the receiving router would mark the state learned from the restarting router as stale and start a timer, the value of which is the lower of the Reconnect Timeout from the FT TLV exchanged at the time of initialization and a configurable local timer called the Neighbour Liveness Timer. When the timer expires, any remaining stale state is purged.

7.5   RSVP WITH TRAFFIC ENGINEERING EXTENSIONS (RSVP-TE)

RSVP is a protocol that was designed for an entirely different purpose—the reservation of resources across a network for a single flow between a source host and destination host. It was chosen as a candidate for a label distribution protocol because it has the ability to define specific paths across the network and associate resources with those paths. This was particularly attractive as it provides a mechanism for traffic engineering of labelled packets. Extensions were added to RSVP to add the ability to request and assign labels along the chosen path. By default, RSVP-TE follows the same path as selected by the IGP. However, it provides the ability to use an alternative path that is based on another set of criteria and explicitly signal that path from ingress to egress of the MPLS network. This is the basis of traffic engineering using MPLS.

The extra information, upon which these traffic engineering decisions are made, is carried in extensions to the IGP (OSPF or IS-IS) and is stored in the traffic engineering database (TED).

7.5.1 *RSVP-TE GRACEFUL RESTART*

The purpose of the RSVP-TE graceful restart mechanism is similar to that of the LDP restart mechanism. The whole point is to avoid the recalculation of the forwarding path during the restart of the control plane of an LSR. The recalculation of an LSP impacts all the LSRs in both the old and the new path between ingress and egress. In a large network, there can be tens or even hundreds of LSPs passing through a single LSR. By maintaining the forwarding state across the restart of the control plane, this widespread disruption is avoided, thus reducing the burden upon the processing power of LSRs.

RSVP restart shares a lot of common features with IS-IS restart including the use of a Restart Timeout and a Recovery Time field in the Hello messages. In the case of RSVP, this is carried in the Restart_Cap TLV. These extensions are described fully in RFC 3473.

7.5.2 *OSPF WITH TRAFFIC ENGINEERING EXTENSIONS (OSPF-TE)*

OSPF has a particular group of LSA types, which are able to carry information with no inherent meaning to OSPF itself. These are known as opaque LSAs. There are three types of opaque LSA. Type 9 LSAs are flooded only on a single link, Type 10 LSAs are flooded only within a single area, and Type 11 LSAs are flooded throughout the entire autonomous system. Type 10 LSAs were chosen to carry traffic engineering information. A new type 10 LSA called the traffic engineering LSA was created explicitly for this purpose. This currently constrains traffic engineering to a single OSPF area. Work is currently underway to develop systems whereby multi-area and even multi-domain traffic engineering can be implemented.

The information carried in the type 10 LSAs is used to populate the TED. The information in the TED is used by the CSPF process to calculate the shortest path, which also meets all the criteria mandated for a particular LSP.

7.5.3 *IS-IS WITH TRAFFIC ENGINEERING EXTENSIONS (IS-IS-TE)*

IS-IS is inherently extensible so, rather than using extant features of IS-IS, new TLVs were created to carry the TE information. The TE information is, as with OSPF, used to populate the TED, which is used by the CSPF process to calculate the path of the LSP. The same extended information types are carried by IS-IS as are carried with OSPF.

7.6 FAST REROUTE

Fast reroute is a term used to describe two mechanisms whereby alternate paths are pre-signalled so that, should a path become unviable, traffic can be transferred onto a new path with minimal delay. The two mechanisms actually protect against different types of failure and can be characterized as either link protection or node protection.

As the name suggests, link protection protects against the failure of a single link by creating a bypass LSP between the two adjacent LSRs either side of the failed link (see Figure 7.3). In the event of a failure, all the affected LSPs are tunnelled through the bypass LSP using label stacking. This relies upon the use of labels, which are unique within the entire chassis (rather than unique on the interface). One major benefit of this is the fact that the labels for the inner LSP remain unchanged. This minimizes the signalling and computational requirements to achieve this form of bypass. In a large network with many nodes this reduction in signalling overhead can be significant.

Unfortunately, link protection is of no value if an LSR totally fails. On the failure of the LSR, all links from that LSR would go down. The LSR at the ingress to each failed link would identify the failure and attempt to tunnel the failed LSPs through an alternate path to the failed LSR (the next hop). Since that LSR is down, the 'tunnel' LSR would also be down and so fast reroute would fail. At this point, all the affected LSPs would be torn down and resignalled end to end. In a large network, the failure of a crucial LSR could cause a massive burst of signalling and path computation.

Node protection protects against the failure of an entire LSR or multiple links associated with a single LSR by creating a bypass LSP between the two LSRs adjacent to the failed LSR (see Figure 7.4). This requires a pre-signalled, alternate LSP for each affected LSP and, upon the detection of the failure, all the outbound packets will have a different label than they would have had. This is much more flexible in the sense that it is possible to route around multiple adjacent failures. However, the signalling and computational overhead of this mechanism are significantly greater than that for link protection as described above.

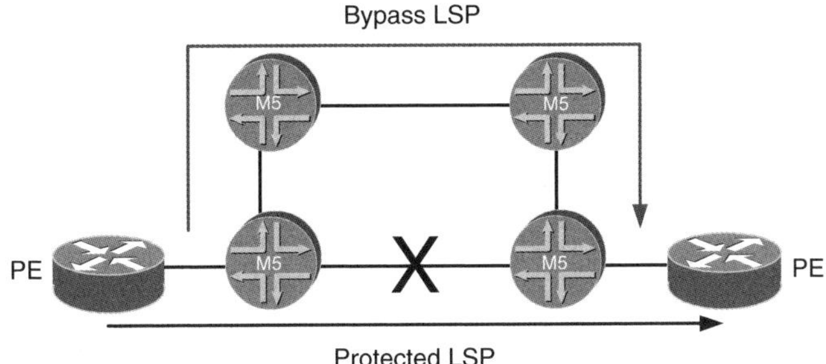

Figure 7.3 MPLS fast reroute, link protection

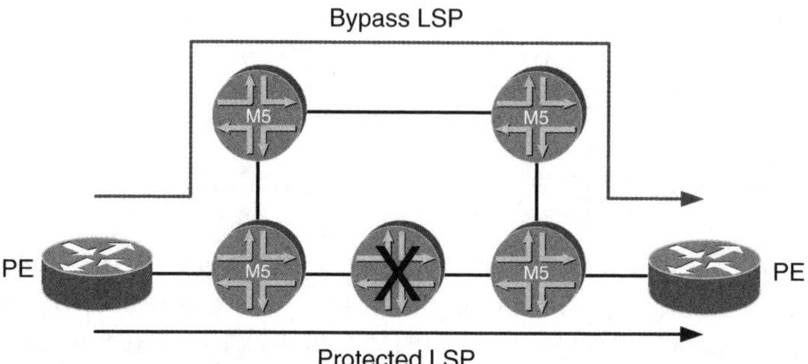

Figure 7.4   MPLS fast reroute, node protection

In addition to the extra bypass path to avoid a failed node, node protection also creates a bypass LSP to avoid a failed single node. In large networks, this can require enormous amounts of state. However, in comparison to link protection, the impact of the failure of nodes protected by node protection will cause minimal signalling and computational overhead.

In general, the functional benefits of node protection outweigh the disadvantages associated with the extra state required. In the end, you have to decide whether the failure of an entire node is sufficiently likely to make the disadvantages of link protection under those circumstances more significant than the benefits of vastly reduced state associated with link protection.

The choice of protection mode will depend upon what has been implemented by your chosen vendors. Cisco and Juniper have both implemented both mechanisms. However, if you have other vendors' equipment in your network, it is important to check their capabilities.

7.7   INTEGRATING ATM AND IP NETWORKS

Since both ATM switches and IP routers can be LSRs, MPLS provides the basis on which this integration can occur. It is generally held to be a good thing that IP networks and ATM networks be integrated with a single control plane and data plane. The main benefits are evident to operators with existing separate ATM and IP networks who wish to operate all the services on a single converged network. By using MPLS, a network provider can significantly reduce the impact of the migration. Equally importantly, it can change the period over which the migration is paid for. For businesses today, this can be an important issue. If new equipment can be purchased over a longer period, then this can make a significant difference to the profitability of the network in the earlier stages of deployment.

ATM switches utilize the VPI/VCI field to carry the label information and use their standard signalling mechanisms for exchanging label information. ATM LSRs also have support LDP as a label distribution protocol. This is the basis on which an LSP can be

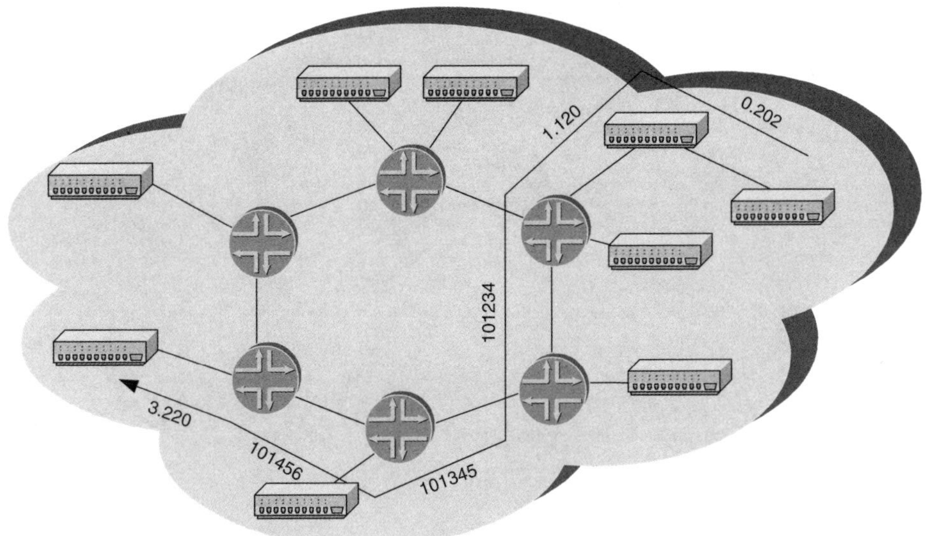

Figure 7.5   Integrating MPLS and ATM networks

built end to end crossing both ATM and IP networks (see Figure 7.5). In so doing, a single 'transport' is created, which is independent of the underlying media.

7.8 GENERALIZED MPLS (GMPLS)

Generalized MPLS takes the concept of MPLS and applies it to various types of Layer 1, Layer 2 or Layer 3 networks. GMPLS devices can have interfaces of the following types:

- fibre-switched capable (FSC);

- lambda-switched capable (LSC);

- time-division-multiplexing switched capable (TSC);

- packet-switched capable (PSC).

Within the GMPLS framework, LSPs must start and finish on links of the same capability. However, the LSP can transit any type of link within the network.

The implication of these extensions is that LSPs can now be built across a network and that devices at Layer 1, Layer 2 and Layer 3 can all participate in the decision how to route the LSPs. Thus, Optical Cross Connects (OXC) and SONET/SDH Add-Drop Multiplexers (ADM) in the core of operators' transport networks can take part in the establishment of LSPs (see Figure 7.6).

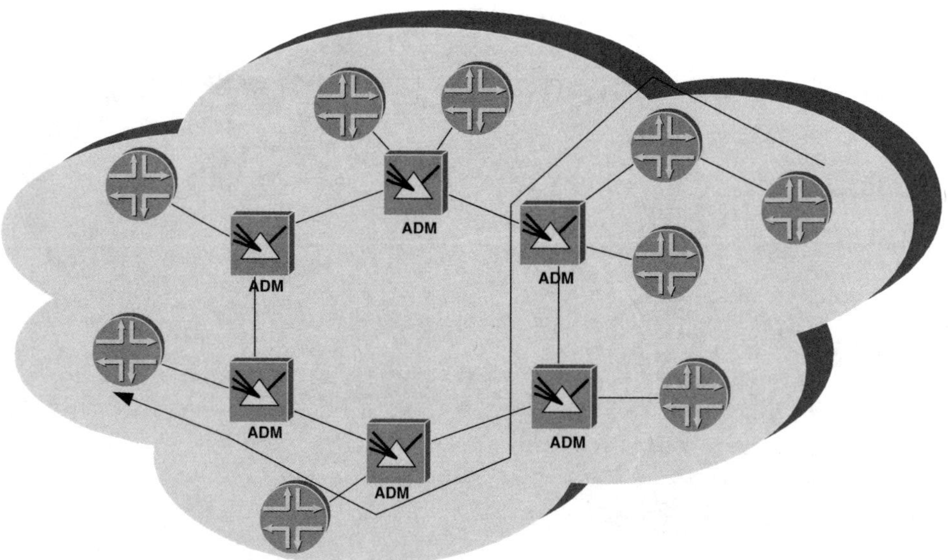

Figure 7.6 Integrating MPLS and Optical switches with GMPLS

GMPLS differs from traditional MPLS in the following ways:

- GMPLS requires a separate Data Control Network over which control messages can be sent. RSVP-TE is an IP-based protocol so cannot be carried natively over the data plane (as is the case with traditional MPLS).

- RSVP-TE has been further extended to allow the LSR to request labels of the form applicable to the particular link (e.g. a wavelength, a fiber, or a time-slot).

- GMPLS can create bi-directional LSPs. Unlike traditional MPLS, in which LSPs are strictly unidirectional, non-packet LSPs in GMPLS can be signalled to be capable of carrying data in both directions over a single path.

GMPLS requires that RSVP-TE and the Link Management Protocol (LMP) for the particular type of link interact very closely. When establishing an LSP, the following steps are required.

1. The LMP informs RSVP-TE of the following elements:

   (a) the Traffic Engineering Link (an interface or sub-interface based on time-slot, wavelength, etc.);

   (b) resources available on the TE Link;

   (c) the control peer.

2. GMPLS derives the specific attributes of the new LSP from the configuration of the node. It requests that RSVP-TE signal one or more specific paths described by the TE Link addresses.

3. RSVP-TE determines the adjacent TE Link, the associated control adjacency and control channel, and the transmission parameters (e.g. IP address). It requests that the LMP allocate the required resources from the TE Link. If the LMP is able to allocate the required resource, then a label is allocated. RSVP-TE then sends a PathResv message towards the destination router.

4. When the PathResv message reaches the destination router, the LMP tries to allocate the requested resources. If it is successful, then RSVP-TE transmits a ResvMsg back towards the source router.

5. Once the ResvMsg reaches the source router, the LSP is established.

As with traditional MPLS, a link state protocol is used to carry the traffic engineering information throughout the network. GMPLS uses OSPF-TE with added extensions to populate the TED, in addition to those used in traditional MPLS networks. The new information is carried in a TLV called the interface switching capability descriptor. The new TLV contains the following attributes:

- signal type (minimum LSP bandwidth);

- encoding type;

- switching type.

GMPLS runs the constrained shortest path first (CSPF) process in exactly the same way as traditional MPLS. The only difference is that it takes the additional constraints into account when calculating the constrained path.

The output of the CSPF algorithm must be a strict path. GMPLS is incapable of using a loose path.

8

Virtual Private Networks (VPNs)

Traditionally, organizations wishing to build a communications network between various remote sites were forced to build a private network from leased lines. This was an expensive and somewhat inflexible method of providing communications between sites. However, then there was no real alternative. Now, there is a viable alternative. Virtual Private Networks are intended to provide a means to provision connections between remote sites with the same characteristics as private leased lines but at a fraction of the price to the customer.

The first virtual private networks were based on Frame Relay and ATM. These two transport media permit carriers to provide virtual circuits between customer sites over a shared core infrastructure. It was possible for carriers to flexibly provision services allocating bandwidth to individual virtual circuits with fine granularity and to allow customers to send more traffic than was guaranteed when the network was lightly loaded.

However, as carriers also became ISPs, they found they were operating several networks in parallel. They wanted to reduce the administrative and operational costs of operating all their networks by converging all their services onto a single IP-based network. If such a move was to work, then VPNs would be one of the main services that would have to be transferred from ATM and Frame Relay onto the new converged network. Below, we look at the various mechanisms used to implement VPNs over an IP-based core.

8.1 VPNS AT LAYER 3

8.1.1 LAYER 3 VPN (RFC 2547BIS)

L3VPN is a mechanism whereby a provider can create a virtual private network at Layer 3. In an L3VPN, the CE 'peers' with and exchanges routing information with the Provider

Designing and Developing Scalable IP Networks   G. Davies
© 2004 Guy Davies   ISBN: 0-470-86739-6

Edge LSR (PE). The PE routers on the service provider's network have knowledge of the customers' routing information. The routing information exchange between CE and PE can achieved using one of a number of protocols. Those most commonly implemented are RIP, OSPF and BGP. It is also possible to use static routing on the PE and a static default on the CE. This information is redistributed at the PE into multiprotocol BGP and is exchanged with all other PEs in the same VPN. The label information is carried in extended BGP communities (draft-idr-bgp-ext-communities-06.txt).

Each PE must maintain a virtual routing and forwarding table (VRF) for each of the VPNs connected to any PE in the entire network. This VRF contains all the routing information for all sites in the VPN. The consequence of this is that all the PEs will contain the complete routing information for all sites in all VPNs. Herein lies the first major scaling challenge with L3VPN. As the total number of VPNs grows, so does the requirement placed upon all the PEs, even if they are not attached to any sites in the new VPN. In Figure 8.1 there are four PEs, each with one or more of the Red, Green and Blue VPNs attached. However, all four PEs must maintain routing state for all three VPNs, even though three of the four have no need for all that state. In addition, as the number of prefixes in each VPN grows, that growth is spread to all PEs, again irrespective of the need for that PE to maintain those routes.

One mechanism to reduce the burden on the PEs is to use route reflectors. If you use route reflectors, only the reflectors need to maintain state for all VPNs. Each PE (assuming it is not one of the route reflectors) only needs to maintain state for those VPNs of which its adjacent CEs are members. In Figure 8.2, the P1 router from the previous figure has been configured as a route reflector and each of the PEs only has a single BGP neighbour. Only RR and PE2 require all the state. Each of the other PEs has a reduced load.

Cisco and Juniper have slightly different recommendations on the best practice with respect to route reflection. Cisco strongly recommends installing separate devices attached in a central location in the core of the network specifically to perform the route reflection function. These devices do not need huge forwarding capabilities since they are not in the forwarding path of any major flows. The only data going to them is routing

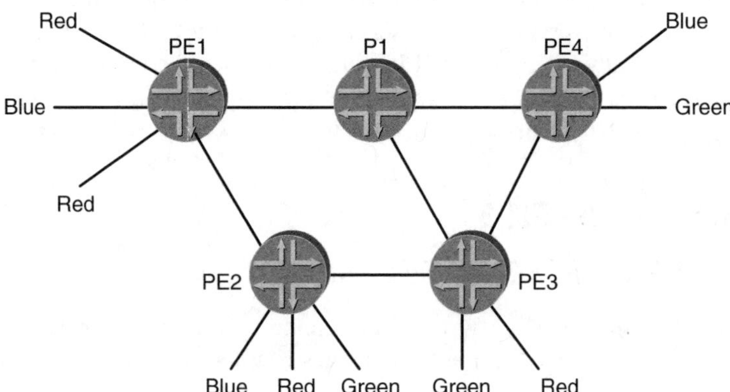

Figure 8.1 MPLS Layer 3 VPN example

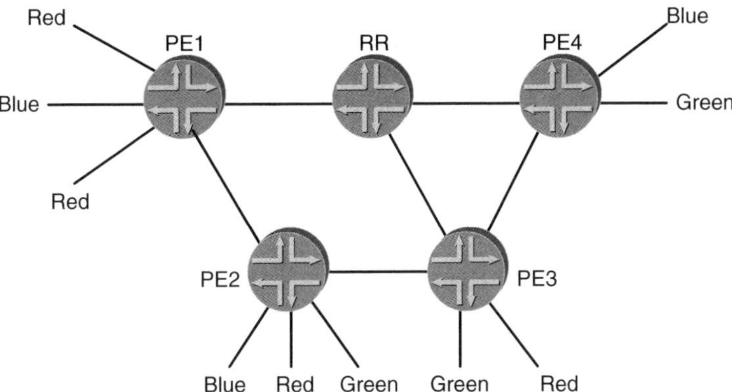

Figure 8.2 MPLS Layer 3 VPN example using route reflection

protocol updates. However, in order to maintain and update the vast amount of state, they do require very large amounts of memory and powerful processors. On the other hand, Juniper generally suggests that it is unnecessary to use a separate device with the inherent extra financial and administrative cost. Instead, Juniper suggests using existing core devices since their core routers all use powerful processors and all have lots of memory. In addition, their architecture suffers minimal degradation in performance as services are added.

As you can see, L3VPN can be somewhat complex. Provisioning new VPNs can be complex and so can adding new sites into existing VPNs. However, the complexity of provisioning is overshadowed by the complexity of troubleshooting these networks. Label stacking means that the PEs must push multiple labels (one for the VPN and one for the path across the network). If, in addition, you are also tunnelling LDP through RSVP-TE LSPs in your core, then you could end up with three levels of label stacking. When there is a failure, it can be immensely difficult to identify which LSR should be advertising labels to which other LSRs using which label distribution protocol. Scaling networks running L3VPN is indeed a major challenge.

8.1.2 *GENERIC ROUTER ENCAPSULATION (GRE)*

GRE is a mechanism for the encapsulation of arbitrary packets over an existing IP network. GRE does not require any MPLS functionality so it can be used to build tunnels through networks without MPLS functionality. This removes the requirement for the P routers in a carrier's network to have any MPLS functionality at all since all MPLS frames are encapsulated inside the GRE tunnel right across the core. Therefore, it has been proposed that the outer MPLS label of a L3VPN could be replaced by a GRE tunnel.

The use of GRE (actually IP+GRE) encapsulated MPLS frames means that the VPN might be more vulnerable to the spoofing of packets into the VPN than if the outer tunnel were MPLS-based. In order to protect against the spoofing of the outer tunnel, it would be necessary to apply filters at every border router (assuming that no spoofed packets will be sourced within the core). This can be administratively intensive. In addition, it is almost impossible to maintain control over the filters in an environment where VPNs traverse multiple carriers' networks.

8.1.3 IPSEC

As described above for GRE, it is similarly possible to encapsulate the inner MPLS label in IPsec. This similarly allows the VPN to be transported across non-MPLS capable portions of a network. However, the use of IPsec as the encapsulation mechanism dramatically reduces the ability to introduce spoofed packets. This enhanced security comes at the cost of reduced efficiency (i.e. IPsec imposes a greater overhead than GRE).

8.2 VPNS AT LAYER 2

8.2.1 CIRCUIT CROSS-CONNECT (CCC)

CCC is a Juniper Networks proprietary mechanism for connecting two circuits over an MPLS core. The circuit at each end of the cross connect can be a Layer 2 interface or an LSP. It predates all the standardized mechanisms (e.g. draft-martini). CCC supports the connection of two interfaces using identical encapsulation. All packets arriving on a particular circuit are simply encapsulated within MPLS frames, transported to the far end PE, decapsulated and transmitted onto the relevant circuit. The Layer 2 frames remain unchanged with the possible exception of the Layer 2 address fields in the headers. Since the complete Layer 2 frame is encapsulated in MPLS, it is essential that the Layer 2 encapsulation is identical at both ends.

Unlike LSPs, cross-connects are bidirectional. Since each cross-connect requires a unique LSP end to end in each direction, there will be a maximum of i LSPs in transit at any point in the network (i is the total number of logical interfaces attached to VPN sites throughout the network and each logical interface can be connected to exactly one other logical interface). This creates a significant burden in terms of state required to maintain all the LSPs. It is particularly inefficient in the presence of a failure since so many LSPs must be torn down, recalculated and the new paths signalled, even though there is likely to be a small number of groups containing potentially large numbers of LSPs where the recalculation for all LSPs in a single group redirects them all onto an identical new path.

In addition to connecting interfaces to interfaces, it is possible to use CCC to connect interfaces to LSPs and to connect two LSPs together. The latter function is known as LSP stitching and is particularly useful for connecting LSPs in different domains without joining the signalling. This was used by some operators to provide connectivity between traffic-engineered LSPs in different traffic engineering domains (e.g. in a multi-area autonomous system) before any standardized mechanism for inter-area traffic engineering was available.

8.2.2   TRANSLATIONAL CROSS-CONNECT (TCC)

TCC is a Juniper Networks proprietary extension of CCC that allows the interfaces at each end of the VPN to use different encapsulation (e.g. Ethernet frames at one end and PPP frames at the other end). It suffers from the same limitations as CCC with respect to the requirement for an LSP from end to end for every 'circuit'.

Translating from one Layer 2 encapsulation to another introduces some new constraints. Due to the requirement to replace the Layer 2 headers at the egress, it is necessary to create a complete encapsulation scheme for each Layer 3 transport. Since this is a Juniper Networks proprietary mechanism and Juniper Networks only supports IP at Layer 3 that is not a big issue. However, it is worth remembering this constraint (honestly).

Another potentially big issue is that of Address Resolution Protocol (ARP). Since this is not a feature of point-to-point protocols like PPP and Cisco HDLC, it is necessary for the PE attached to the Ethernet end to generate 'proxy ARP' responses for those devices at the far end of the link.

8.2.3   MARTINI (LAYER 2 CIRCUITS)

'Martini' is not just one draft but actually a set of drafts describing a mechanism for transporting Layer 2 PDUs over MPLS and a set of encapsulation mechanisms, one for each different type of Layer 2 PDU. Each tunnel requires one label to identify the egress port or channel on the egress routers and a second label to identify the egress PE.

Unlike CCC and TCC, Martini tunnels between the same pair of PEs can all share a single transport LSP between those PEs. The egress interface is identified by the inner label in the stack. This dramatically reduces the impact when a failure occurs. There is a maximum of 2n(n-1) LSPs traversing a single P LSR where n is the number of PEs to which VPN sites are connected. Assuming that there is a comparatively small number of PEs and each PE has a comparatively large number of attached sites, this will impose less of a burden. For example, if there are 5 PEs, each with 50 connected sites you will see the following numbers of LSPs:

$$\begin{array}{ll} \text{CCC/TCC:} & 2 \times 50 \times 49 = 4900 \\ \text{Martini:} & 5 \times 50 = 250 \end{array}$$

Martini tunnels are signalled using directed LDP. In directed LDP, the address of the LDP 'neighbour' is explicitly configured. This neighbour is reached through a standard LSP that can theoretically be signalled using any label distribution protocol. This use of a transport LSP and a label to identify the logical interface means that it is necessary to stack at least two labels in the MPLS header.

The standardized encapsulations for each of the Layer 2 protocols are defined in a range of Internet drafts and RFCs. In some cases these are identical to the encapsulation used for that Layer 2 protocol in CCC.

By using multiple logical interfaces on the physical interfaces connecting a CE to a PE, it is possible to create a full mesh between all the sites or just a partial mesh.

8.2.4   VIRTUAL PRIVATE WIRE SERVICE (VPWS)

VPWS is a generic term for a point-to-point Layer 2 VPN. There are a number of specific mechanisms for the creation of virtual private wires. Particularly, there are several alternative mechanisms for signalling these virtual private wires.

8.2.4.1 BGP-Signalled Layer 2 VPNs

Juniper's preferred mechanism is known as L2VPN (a rather generic name now given the array of different Layer 2 VPN technologies). It uses multiprotocol BGP to exchange label information between PEs in a very similar mechanism to that developed for L3VPNs in RFC 2547bis (see above). The NLRI formats and the specifics of how BGP carries the NLRI are defined in draft-kompella-ppvpn-12vpn-xx.txt. However, the discovery mechanism is more generically defined in draft-ietf-13vpn-bgpvpn-auto-xx.txt. You may have noticed that this draft is actually 'owned' by the 13VPN working group rather than the 12VPN working group. This simply demonstrates the similarity in the requirements of the two mechanisms.

L2VPN is an interesting paradox. It shares some features with L3VPN, which is difficult to scale as described above. However, the most constrictive elements of L3VPN are not shared by L2VPN. Since this is a Layer 2 service, the PE routers carry none of the customers' routes; they only carry information regarding the label associated with a particular egress interface. Since this requires only two entries for each point-to-point link in the VPN (one for each direction), it scales linearly with the growth of the number of links in the VPN.

Despite the amount of state growing at a linear rate, the actual number of LSPs through the core is limited to 2n(n-1) (as with Martini), where n is the number of PEs in the network. Each PE also has to maintain state for the LSPs for which it is the ingress and egress device but this information is entirely hidden from the P LSRs in the network. The number of prefixes maintained only on the PE LSRs grows linearly with the number of sites connected to the PE LSRs.

The benefits of this mechanism include the ability to 'pre-provision' channels to existing sites in a VPN so that new sites can be added to the VPN without having to touch any of the PEs except the one to which the new site is attached. The problem with this is that it is necessary to assign a base label and a maximum number of labels. Unless you are a perfect fortuneteller, this is likely to lead to either wasting usable space in the label range or having to make a disruptive change to the configuration of the PE to increase the available space.

In common with Martini, it is possible to set up full or partial meshes or hub and spoke structures for the VPN.

L2VPN is considered by some to solve some of the scaling issues associated with the administration of Martini because it provides mechanisms to semi-automate the discovery of new sites in an existing VPN. This contrasts against the requirement with Martini to manually configure both ends of each point-to-point tunnel. However, the proponents of Martini suggest that overloading the BGP protocol to carry more and more NLRI when you have a perfectly functional label distribution protocol (LDP) is unnecessary. The great thing for network operators is that both groups have developed implementations of their drafts. Martini is more widely implemented than L2VPN so if you are looking for interoperability in a multi-vendor environment, Martini may be the easier choice. If you have only one brand of PE, but another brand of P LSR, then running L2VPN is a reasonable alternative.

8.2.4.2 RADIUS Signalled Layer 2 VPNs

An alternative proposal to the use of BGP as a signalling protocol for auto-discovery of VPNs is to use RADIUS. This is described in draft-heinanen-radius-pe-discovery-XX.txt.

In order for a site to connect to a VPN it is necessary to create an entry in the RADIUS database containing three items:

- The site name (e.g. [providerP/]siteX@blue.yourdomain.com). This is analogous to the username. The providerP/prefix is only required if the PE is not owned by the administrator of the VPN (provider P).

- The password (e.g. siteXpassword).

- The VPN identifier (e.g. blue.yourdomain.com).

When the CE tries to connect, the PE sends a RADIUS access request to the RADIUS server. If the site is known to the server, then access is granted and the RADIUS server informs the PE of the VPN ID to which the site should be connected and a list of PEs that have attached sites for this VPN. The RADIUS server records the following information.

- the name of the CE;

- the IP address of the PE;

- the VPN ID;

- the timestamp of the most recent successful authentication of this site.

There are potential scalability benefits to using RADIUS. RADIUS servers are operated by many network operators already for AAA of customers (dialup, DSL, wireless connectivity). RADIUS has been widely deployed in environments with many tens or even hundreds of thousands of users who connect and disconnect fairly rapidly. VPN sites are generally likely to be comparatively stable. Therefore, it is clear that it is possible to have hundreds of thousands of VPN sites being connected and authenticated using RADIUS.

The widespread deployment of RADIUS also means that the administrative cost of implementing a RADIUS-based solution should be low.

One of the major disadvantages of this mechanism is that it relies upon an external device to mediate access to the VPN. With BGP-based auto-discovery, each PE is responsible for mediating local access to the VPN. Thus, if the PE goes down, it is only access via that PE that is impacted, not all future access via all PEs (until the RADIUS server is fixed).

8.2.5   VIRTUAL PRIVATE LAN SERVICE (VPLS)

As the name suggests, this is a mechanism to provide a virtual LAN between remote sites giving apparently switched LAN connectivity between those sites. It is implemented as a set of point-to-multipoint VPNs.

A VPLS CE can be a Layer 2 Ethernet switch or a Layer 3 router with an Ethernet interface. Given that VPLS is emulating a LAN, it is only necessary to have Ethernet encapsulation. However, there are some new challenges when implementing a virtual private LAN over a large network with many PEs. In particular, it is necessary for the CE and PE to learn Ethernet MAC information so that it is possible to correctly direct packets through the network.

Each CE in the VPLS instance connects to a PE and, just as any other edge device, it sends Layer 2 frames onto the network. If the destination is on the same subnet but not on the same segment, the bridge (in this case, the bridge is the VPLS PE) receives the packet and recognizes that it is not for a local host. On receipt of the frame, the PE must decide if it knows the destination for the packet. If it does, the packet is sent directly to the appropriate remote PE. If the PE does not know the destination for the packet, the packet is broadcast to all other PEs within the VPLS instance. Each PE then forwards the packet to its associated CEs. The CE, to which the destination is connected, forwards the packet. All other CEs simply drop the packet.

Figure 8.3 shows a network with four PE LSRs and four CE switches. There are workstations attached to CE1 and CE3 that wish to communicate. Let us assume that no

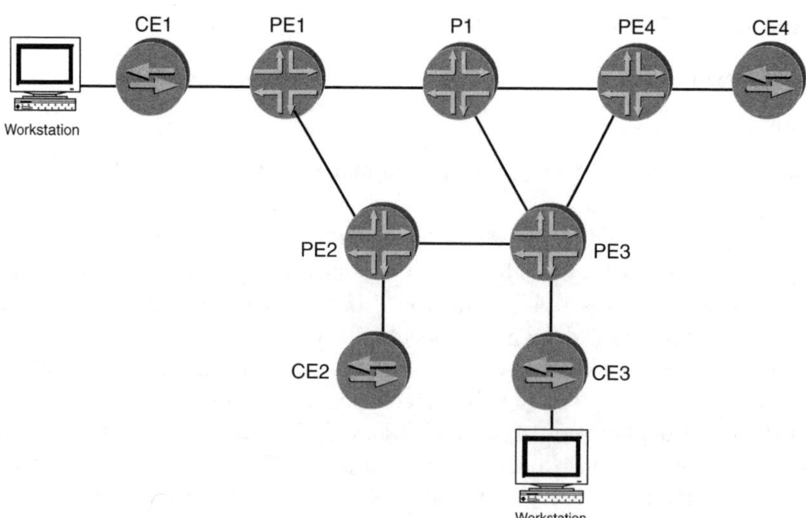

Figure 8.3   VPLS example

Ethernet MAC addresses have been learned by any of the PE LSRs in the network, that the CEs know the MAC addresses of their attached workstations and that workstation 1 (attached to CE1) wishes to send a packet to workstation 3 (attached to CE3).

Workstation 1 transmits the packet to CE1. CE1 now knows that workstation 1's Ethernet MAC address is reached via the port over which the packet was received. Since CE1 does not know how to reach workstation 3, it broadcasts the packet on all ports. PE1 receives the packet. It does not know how to reach workstation 3's MAC, so it also broadcasts the packet to each of the known PEs. PE2, PE3 and PE4 each receive a copy of the packet. None of them knows the path to the MAC for workstation 3 so they broadcast the packet. The PE LSRs transmit to CE2, CE3 and CE4 respectively. Neither CE2 nor CE4 have the MAC address for workstation 3, so they silently drop the packet. CE3, however, knows that workstation 3 is attached directly to it and transmits the packet to the destination. In the process of transmitting the packet throughout the network, each of the PE LSRs and CE switches now knows the path back to workstation 1's MAC address.

Workstation 3 then replies to the received packet. It transmits to CE3. CE3 knows that workstation 1 is reached via PE3. PE3 knows that it has to send the packet back to PE1. PE1 knows that workstation 1 is connected via CE1. CE1 has a direct connection to workstation 1. In addition, CE3, PE3, PE1 and CE1 now know the path back to workstation 3's MAC address. Note that because the packet from workstation 3 was not broadcast, CE2, PE2, PE4 and CE4 do not have any information about the MAC address of workstation 3. This mechanism for learning the MAC addresses of attached devices is analogous to the behaviour of a standard learning bridge.

VPLS on Juniper routers requires the use of a Tunnel Services PIC (or an Adaptive Services PIC) because two labels must be pushed and popped in a single device.

8.2.6   LAYER 2 TUNNELLING PROTOCOL (L2TP)

L2TP is a protocol that originally was designed to provide a mechanism to transport PPP frames across an intervening Layer 3 network. The original protocol is defined in RFC 2661. Subsequent extensions to the protocol have facilitated the use of L2TP to carry other Layer 2 frames. It is clear that L2TP could potentially be used as a means for providing point-to-point VPNs.

Many believe that GRE and L2TP provide more efficient mechanisms to transport the data in VPNs and, when used in conjunction with auto-discovery mechanisms such as the RADIUS-based system described earlier in this chapter, can provide all the flexibility and scalability of an MPLS-based VPN.

9

Class of Service and Quality of Service

From the earliest days of the Internet, traffic has been transported on a best efforts basis. For a long time, this was perfectly satisfactory. However, at a time when SPs can charge more for premium services, best effort delivery is no longer sufficient.

Traditionally, service providers operated separate networks for their data, video and voice services, each of which placed different quality demands upon the networks over which they were operated. However, the desire for financial savings based on building, operating and maintaining only one network rather than several has driven service providers towards convergence of those networks. This has led to vastly more stringent demands upon their IP networks. Various features exist within IP, MPLS and the underlying transport media to support Quality of Service (QoS).

9.1 DESIGN AND ARCHITECTURAL ISSUES OF CoS/QoS

As stated above, Class of Service (CoS) and QoS place new demands upon the IP and MPLS infrastructures operated by service providers. These are particularly relevant to the scaling of networks. The extra control traffic and state required to maintain the quality of service across a network can place extra demands on the processors and memory of all network devices. In addition, the management of queues and scheduling can be especially complex. Many products have a constrained number of queues per port, several having just four queues. While this can be an issue, particularly in environments where per-subscriber class of service implementations are required, in core environments it is uncommon to

Designing and Developing Scalable IP Networks   G. Davies
© 2004 Guy Davies   ISBN: 0-470-86739-6

require vast numbers of queues. In this environment, CoS and QoS are applied to highly aggregated flows of data.

9.2 CoS/QoS FUNCTIONAL ELEMENTS

CoS and QoS are pretty generic terms. However, there are a number of functional elements required for a full CoS/QoS implementation.

9.2.1 CLASSIFICATION

In order for one class of traffic to be treated preferentially over a different class of traffic, it is clearly necessary to be able to identify traffic that falls into each of the following classes:

- *Ingress classification.* Classification is generally carried out as close to the edge of the network as possible. Ideally, this would be right out at the host sourcing the packet. However, allowing customers to set the class of a particular packet would potentially cause problems if not adequately monitored. Consider the situation where you offer four service levels to your customers (e.g. gold, silver, bronze and best efforts) and that gold obtains better service than silver, which obtains better service than bronze, which obtains better service than best efforts. If you leave it entirely up to your customers and apply no monitoring or reclassification at your borders, it is highly likely that your customers will associate all packets with the gold class of service.

- *Per-hop classification.* Despite the fact that a classification is applied at the ingress to the network, it is still necessary to perform a classification function as each packet arrives at each network device. If the ingress classification has been recorded in the Layer 2 or Layer 3 headers of the packet, then the process of classification is simple. If the classification has not been marked, then a complete reclassification exactly as at the ingress to the network must be completed at each step.

- *Rewriting on egress.* The default behaviour with all the QoS mechanisms is to leave the CoS/QoS markings exactly as they were when the packet arrived at the device. However, there are a number of situations when it is desirable to modify the marking of the packet before it is transmitted. The obvious example is at the point where the packet is initially classified. The packets arrive with no classification markings. Then, through some arbitrary classification process, they are identified as belonging to a particular class. If you do not want to have to reclassify each packet at each hop throughout the network based on those arbitrary criteria, it is necessary to modify the QoS marking. This allows subsequent devices along the path to trivially identify the class into which packets should be placed.

9.2.2 CONGESTION NOTIFICATION MECHANISMS

Despite all your best efforts, it is almost inevitable that your network will experience congestion at some point, even if it is only on the lower bandwidth links to your customers. It is extremely useful to be able to inform the source and, to a lesser extent, the destination that there is some congestion in the path between them. This notification provides the information based upon which the end hosts can modify their behaviour. Retransmission of lost data can actually exacerbate congestion because the packets that previously could not be transmitted are retransmitted in addition to the new data.

For example, I have seen traffic load on a new circuit drop by as much as 25% when a massively over-provisioned link was substituted for a significantly over-subscribed link. A pair of E2 (8 Mbps) links was struggling hard trying to carry an offered load of 20 Mbps between two routers. When these two links were replaced with a single STM-1 POS link, the total traffic over the link dropped to around 15 Mbps.

9.2.2.1 Implicit Congestion Notification

Several protocols rely upon an acknowledgement (ACK) being sent for each packet received. The loss of a packet, or of the acknowledgement packet, will be noted by the transmitting host and it will send the packet again. However, it is implicit in the loss of the packet that there is congestion somewhere on the path from source to destination or on the return path. Since there is congestion, the transmitting host can slow down (the mechanism used to slow down the transmission is protocol specific). If an ACK is dropped, then the receiver will receive a second copy of the transmitted packet. It can also, therefore, assume that its ACK for that packet was dropped, probably due to congestion in the path back to the source. On receipt of the duplicate packet, the destination can slow down the speed at which it transmits packets back to the source.

9.2.2.2 Explicit Congestion Notification

Implicit congestion notification is effective only in protocols that require ACKs and that adapt their behaviour to the loss of packets. Many protocols simply do not use ACKs so an alternative mechanism is required. They can use explicit congestion notification (ECN) techniques to inform both source and destination of the congestion. These messages are sent to both source and destination by the device on which congestion is experienced.

These notification messages are sent back to the source in packets travelling in the reverse direction. In TCP packets, ECN can be marked using the two high-order bits of the ToS byte. Bit 6 is the ECN capable transport (ECT) bit that is used to identify a source as capable of responding to ECN messages. Bit 7 is the congestion experienced (CE) bit, which is set by the router on which congestion was identified. This is rarely implemented. The DiffServ standard for the use of the ToS byte describes bits 6 and 7 as unused.

ECN is more widely implemented in ATM and Frame Relay switches. Frame Relay uses 3 bits for forward explicit congestion notification (FECN), backward explicit congestion notification (BECN) and discard eligible (DE). FECN and BECN are set by the data communication equipment (DCE) on which congestion occurred to inform data terminating equipment (DTE) of the congestion. FECN indicates that traffic arriving at the DTE experienced congestion. BECN indicates that traffic transmitted by the DTE experienced congestion. DE is set by the source DTE to provide a very crude mechanism to identify lower priority traffic that can be dropped by the DCE in preference to other, nominally higher priority, packets.

In TCP, BECN packets are sent in the ACK packets sent back to the source and on that basis, the source can slow down. However, with UDP packets, there are no inherent replies (no ACKs) so there are no packets in which to set the BECN and the source cannot be informed of the congestion. In order to overcome this, on receipt of the packet with the FECN bit set, the DTE can send a test frame back towards the source with the BECN bit set. The DE bit in these test frames must be cleared to minimize the chance of that test frame being dropped.

9.2.3   *CONGESTION AVOIDANCE MECHANISMS*

The whole reason for QoS and CoS is to make a selection between packets when two or more packets require transmission at the same time but there are insufficient resources to transmit both. This is known as congestion and, while it is possible to deal with congestion after it has happened, it is better to avoid the situation in the first place. Congestion avoidance mechanisms make use of features of the protocols to cause certain arbitrarily selected flows to slow down their transmission speed.

9.2.3.1 Random Early Detect

When left unchecked, an over-subscribed link will tend to see a throughput pattern characterized by the graph in Figure 9.1.

This effect is known as global synchronization. It is caused by the synchronization of the reduction and expansion of TCP window sizes in all (or a significant proportion of) the flows traversing the link. As the utilization approaches 100%, the buffers start to drop packets from the rear of the buffer as there is no room to buffer the packets requiring transmission. This is known as tail dropping and is the worst possible form of dropping because of the inherent lack of control over the process. High-priority packets are dropped with equal preference to low priority packets. As packets are dropped, the TCP sources start to notice that they are not receiving TCP acknowledgements (ACKs) for transmitted packets. This causes them to reduce their TCP window size, thus reducing the volume of traffic being transmitted. As the link becomes even more heavily over-subscribed, many flows are simultaneously affected and they all reduce their window size simultaneously. This causes the link to become dramatically under-utilized. Gradually, all the flows will increase their window size and the utilization approaches 100%. This starts the whole process again

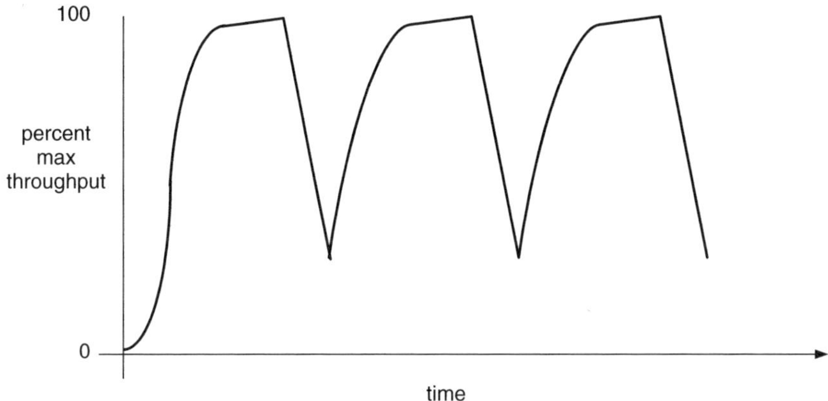

Figure 9.1 Global TCP synchronization

causing a clipped sinusoidal throughput over time. This makes the overall throughput of the link significantly less than could be carried if global synchronization did not occur.

Random Early Detect (a.k.a. Random Early Discard) uses the windowing mechanism of TCP to cause source hosts to slow down. By dropping randomly selected packets, RED prevents the receiver of the TCP flow from receiving the packet and, therefore, prevents the receiver from sending an ACK. When the TCP source does not receive an ACK for the discarded packet, it reduces its window size. This effectively reduces the throughput of the flow.

The random nature of the selection of packets to be discarded means that no single flow is unduly affected. However, the overall impact is to round off the growth of traffic flowing over a particular link (see Figure 9.2). The maximum utilization of the link is slightly below 100% because packets are dropped more aggressively as the utilization approaches 100%.

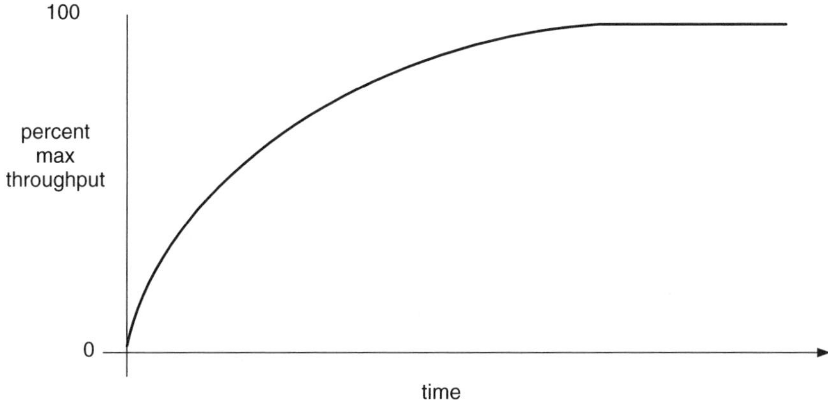

Figure 9.2 Output with RED applied

Most vendors provide mechanisms for the network operator to configure steps or levels at which dropping occurs and the proportion of packets dropped at each level. RED, by default, treats all flows as equal. Thus, packets in your high-priority classes are just as likely to be dropped as those in your low priority classes. Clearly, this is not optimal if a flow is particularly intolerant to packet loss. Therefore, it is necessary to apply different drop profiles to each class of traffic. Generally, two profiles are sufficient (normal and aggressive drop profiles) but there is nothing to prevent each class having a unique drop profile. The main constraint is complexity.

RED profiles are based on a set of mappings of percentage throughput to a drop probability. These create stepped profiles as in Figure 9.3.

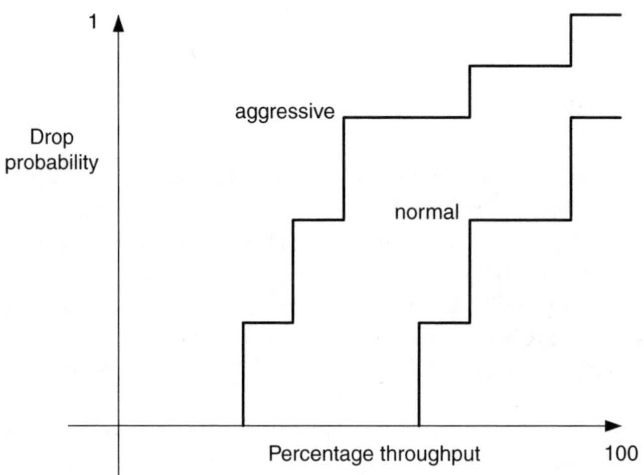

Figure 9.3 Stepped weighted RED parameters

9.2.4 QUEUEING STRATEGIES

Queueing is the basis of the implementation of required egress behaviours. Different vendors provide widely differing numbers of queues on their devices, some using software-based queues (less common today) and others using hardware-based queues.

The primary basis of providing preferential treatment to particular packets is to service different queues at different rates. For example, going back to the earlier example of four services, gold, silver, bronze, and best effort, it might be that the gold queue is served 4 times, silver served 3 times, bronze served twice and best efforts served once (a total of 10 units of service) (see Figure 9.4). This is a somewhat simple example but it demonstrates the point.

9.2.4.1 Fair Queueing

Fair queueing is a mechanism used to provide a somewhat fairer division of transmission resources. Fair queueing ensures that bandwidth is shared equally among the queues on an

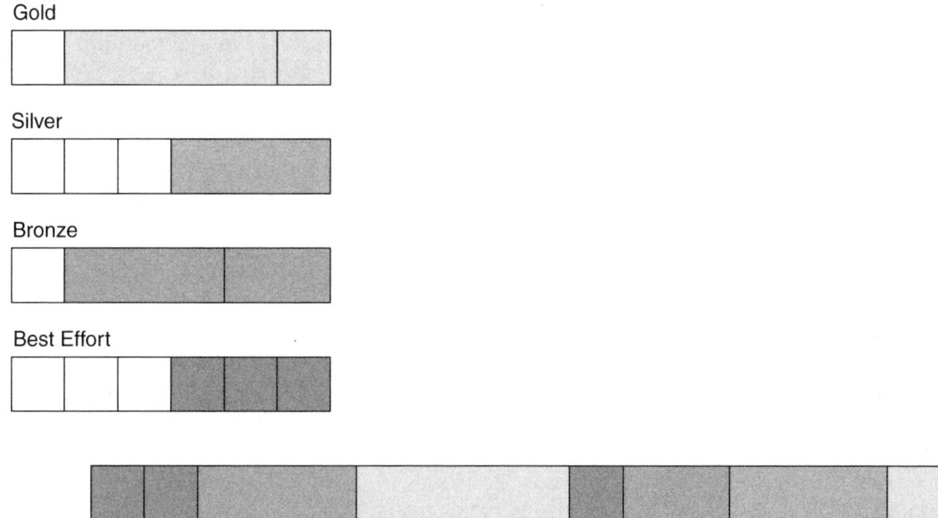

Figure 9.4   Weighted queueing schedulers

approximately per-bit basis, thus ensuring that no queue becomes starved of bandwidth. Each queue is serviced in turn with empty queues receiving no service. Their service is shared equally among the queues with packets to be transmitted. Fair queueing also ensures that low bandwidth flows and high bandwidth flows receive a consistent response and that no flow can consume more than its fair share of output capacity.

FQ does have some limitations, some of which are implementation specific and some are associated with the algorithm itself:

- FQ assumes that all packets are the same size. It is much less effective when packets are of significantly differing sizes. Flows of larger packets are favoured over flows of smaller packets.

- FQ is implicitly fair, i.e. all flows are treated equally. There is no mechanism to favour a high priority flow over a low priority flow.

- FQ is sensitive to the order in which packets arrive. A packet arriving in an empty queue immediately after the queue has just been skipped will not be transmitted until that queue is considered again for service. Depending upon the number of queues, this can introduce a significant delay.

FQ has often been implemented as a software feature. On such devices, this inherently limits the speed at which it can be operated so it is constrained to lower speed interfaces.

9.2.4.2 Weighted Fair Queueing (WFQ)

Despite its name, WFQ is not a direct derivative of FQ but was developed independently. Looking back at Figure 9.4, the gold queue is given 4 credits. It may transmit packets until its credits are all gone or until it does not have enough credits to transmit a whole packet. So, it can transmit the 1 unit packet but, even though it has 3 remaining credits, it cannot transmit the 4 unit packet. Therefore, despite the fact that it is provided with 4 times as many credits as the best efforts queue, it actually ends up transmitting exactly the same amount of data in the first round.

WFQ overcomes some of the limitations of pure FQ:

- WFQ provides a mechanism to favour high-priority flows over low-priority flows. Those flows are put into queues that are serviced proportionately to the level of their priority.

- WFQ supports fairness between flows with variable packet lengths. Supporting fairness in these circumstances significantly increases the complexity.

However, WFQ still suffers from some of the limitations of pure FQ and suffers from some of its own making:

- Due to its computational complexity, WFQ is difficult to apply to high-bandwidth flows. This limits its usefulness to low-bandwidth interfaces. This is further exacerbated by the fact that high-speed interfaces introduce such a low serialization delay that the benefits of WFQ are too small (verging on insignificant) on high-speed interfaces.

- WFQ is generally applied to a small number of queues associated with each interface into which vast numbers of flows are placed. The theoretical benefits of WFQ in preventing the misbehaviour of one flow from affecting other flows are not provided to other flows within the same queue.

- WFQ is still implemented in software. This places constraints upon the scalability of WFQ within a routing device.

On Cisco routers, WFQ is enabled by default on low-speed (2 Mbps and lower) interfaces.

9.2.4.3 Round Robin

This is simply the process of servicing each of the queues in turn. Each queue gets an equal level of service so, in that respect, pure round robin is of very limited value to a network operator trying to provide differentiated QoS. In its purest form, the queues are serviced in a bit-wise fashion (i.e. the first bit is transmitted from each queue in turn). This is the closest it is possible to get to the ideal theoretical scheduling algorithm called Generalized Packet Scheduler (GPS).

9.2.4.4 Weighted Round Robin (WRR)

Weighted round robin provides the means to apply a weight to the rate at which each queue is serviced. This is the simplest mechanism for providing differentiated QoS. As with round robin, in its purest form, weighted round robin is difficult to implement because it would require the fragmentation of packets into blocks proportional in size (in bits) to the weight applied to the queue. This would actually not work because it would not be possible to reassemble the packets at the far end of the link or at the destination host.

9.2.4.5 Weighted Deficit Round Robin (WDRR)

Weighted deficit round robin is somewhat similar to WRR and WFQ. However, it selects the queue to be serviced in a slightly different way to a pure WFQ. If we go back to Figure 9.4, we can see that in the starting position, the gold queue has a 1 unit packet and a 4 unit packet. With WFQ, the gold queue can only transmit one packet because it does not have sufficient 'credit' to transmit its second packet. With WDRR, the second packet is transmitted. However, at the next round, the gold queue has a deficit of 1 before its weight of 4 is applied because it transmitted one more unit than it was due at the first round. So on the second iteration the gold queue is only given 3 units rather than its nominal weight of 4 units.

WDRR has one major benefit over WFQ in that it is easier to implement and, therefore, is easier to apply to very high speed interfaces. On the downside, WDRR cannot protect effectively against a badly behaved flow within a highly aggregated environment. Also, it is not able to provide quite such accurate control over latency and jitter of packets in the queues as can be provided by priority queueing, for example.

9.2.4.6 Priority Queueing

It is sometimes desirable to service one class of traffic to the possible detriment of all other traffic. For example, any traffic requiring low-latency, low-jitter service (e.g. voice or video over IP) might be placed in such a queue. In order to achieve this, one queue can be created into which the high-priority traffic is placed. If this priority queue contains any packets, it is always serviced next. One possible downside of this mechanism is that it is clearly possible to starve all the other queues of service. In order to mitigate this risk, some vendors have provided a configuration mechanism to place an upper limit on the bandwidth used by the priority queue.

As with most things, no one solution is ideal for all situations. Therefore, it is often necessary to implement a combination of the above queueing techniques to provide the broad range of services required by your customers.

9.3 QoS MARKING MECHANISMS

QoS marking mechanisms are used to provide a consistent, arbitrary marker on each packet that can be used by each device in the path (Layer 2 or Layer 3) to identify the class to which that packet belongs and, therefore, put each packet into the relevant queue. This is the means whereby a consistent behaviour can be achieved right throughout your network.

9.3.1 LAYER 2 MARKING

9.3.1.1 802.1p

Ethernet has 3 bits in the MAC layer header, which are allocated for the specification of eight levels of QoS. Their use is defined in the IEEE standard 802.1p. This allows an Ethernet switch to prioritize traffic without having to look into the Layer 3 headers (see below). The details of the mechanism used to implement the required behaviour associated with a specific 3-bit pattern is hardware dependent. However, as with all QoS functions, the treatment of outbound packets is based on queueing.

The queueing mechanisms above can all be applied to the packets transiting Ethernet switches. The 802.1p bits are simply used to classify packets and put them into the relevant queue.

9.3.1.2 ATM

ATM has very highly developed QoS functionality, which was integrated into the protocol from its first development. ATM provides for a number of qualities of service. They are signalled during the set-up of each VC and are strictly enforced by switches in the path. The qualities of service available are described as follows:

- Class A—Constant Bit Rate (CBR) is a stream of data at a constant rate often of a nature, which is intolerant of delay, jitter and cell loss. CBR is defined in the form of a sustained cell rate (SCR) and a peak cell rate (PCR). The SCR is a guaranteed capacity and PCR is a burst capacity, which can be used should the bandwidth be available. An example application of this class is TDM traffic being carried over ATM.

- Class B—Variable Bit Rate—Real Time (VBR-RT) is a stream of data, which is bursty in nature but is intolerant of jitter and cell loss and especially intolerant of delay. An example application of this is video conferencing.

- Class C—Variable Bit Rate—Non-Real Time (VBR-NRT) is a stream of data, which is bursty in nature but is more tolerant of delay. An example application of this is video replay (e.g. training on demand).

- Class D—Available Bit Rate (ABR) and Unspecified Bit Rate (UBR). These are both 'best effort' classes. UBR is subject to cell loss and the discarding of complete packets. ABR provides a minimum cell rate and low cell loss but no other guarantees. Neither ABR nor UBR provide any latency or jitter guarantees. An example application of this is for lower priority data traffic (e.g. best effort Internet traffic).

9.3.1.3 Frame Relay

In common with ATM, Frame Relay has highly developed QoS functionality. Frame Relay also uses the concept of permanent virtual circuits (PVC) and switched virtual circuits (SVC).

Frame Relay uses the concept of Committed Information Rate (CIR) that is analogous to SCR in ATM, and a burst rate that is analogous to PCR in ATM to provide a guaranteed bandwidth and also a maximum rate up to which traffic can burst for a brief period.

9.3.2   LAYER 3 QoS

IP provides one byte in the header of every packet that can be used to identify the Type of Service (ToS) of that packet. The ToS byte is divided up as shown in Figure 9.5.

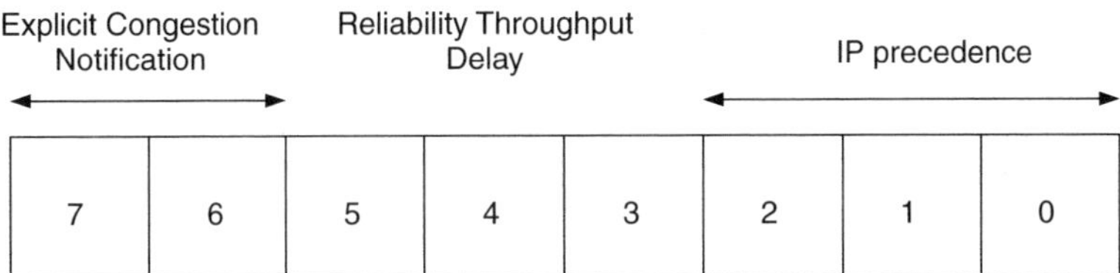

Figure 9.5   The IP Type of Service Byte

9.3.2.1 IP Precedence

The three lower order bits (0–2) of the ToS byte are used to indicate IP precedence. This provides the means to identify eight classes of service. Bit 3 is used to represent normal/low delay, bit 4 to represent normal/high throughput and bit 5 to represent normal/high reliability. Bits 6 and 7 are unused and should be set to zero. Very few applications actually make use of bits 3 to 5 for the purposes for which they were reserved.

The IP precedence bits are possibly the most widely used Layer 3 QoS marking mechanism.

9.3.2.2 Differentiated Services (Diffserv)

Since so few applications make use of bits 3 through 5 of the IP ToS byte, the Differentiated Services working group suggested the use of those three bits in addition to the three low-order bits to provide the ability to identify up to 64 classes of service, known as Diffserv Code Points (DSCP) (see Figure 9.6).

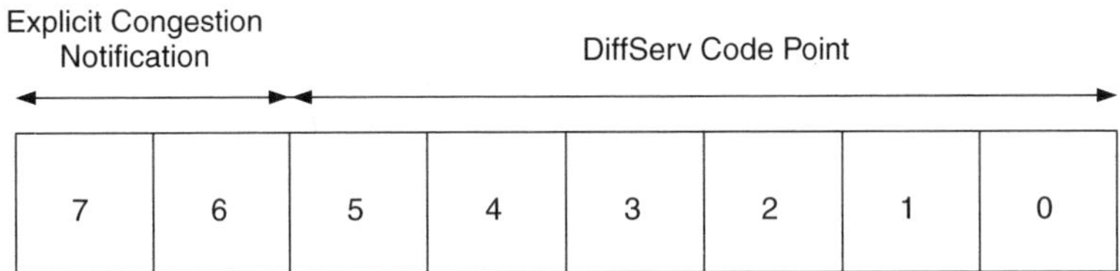

Figure 9.6   The DiffServ Code Point

Diffserv provides a schema for signalling and enforcing QoS from end to end. Diffserv introduces two important concepts. With IP forwarding, the DSCP is evaluated at each hop along the path. At each hop, each class of service is known as a per-hop behaviour (PHB). A behaviour aggregate (BA) represents the sum of all traffic traversing an interface to which the same PHB applies.

9.3.3 MPLS EXP

As with the various devices at Layer 2 and the ToS byte at Layer 3, the EXP bits in the MPLS header are used to identify the queue into which MPLS packets are placed on egress. Also, in common with other marking mechanisms, they provide a persistent mechanism of marking each packet with an identifier that can be used to apply a consistent behaviour from the ingress of the network to the egress of the network.

9.4 INTEGRATING QoS AT LAYER 2, IN IP AND IN MPLS

It is somewhat convenient that there are three bits assigned for QoS in Ethernet frames, IP and MPLS because this makes it possible to create an integrated policy that maps all three QoS mechanisms to each other. As packets pass from one format to another, a consistent behaviour can be achieved.

This can be achieved by the creation of rewrite policies that write the bits in one of the headers based upon the queue into which the packets have been placed. This may not provide a perfect mapping between one set of headers and another. For example, if a device supports four queues, then it might be necessary to place packets marked with two different classes into a single queue, for example, see Table 9.1.

In this case, a rewrite policy cannot differentiate between packets marked 000 and packets marked 001. Since all such marked packets are put into queue 0, they can only

Table 9.1 Mappings from IP ToS bits settings to output queuers using four output queues

Class of Service	QoS marking (assumes 3 bits)	Queue
Best Effort	000	0
Best Effort—High Priority	001	0
Assured Delivery	010	1
Assured Delivery—High Priority	011	1
Low Latency	100	2
Low Latency—Low Jitter	101	2
Network Management	110	3
Network Control	111	3

be rewritten with one value. So, if a rewrite of the 802.1p bits is required, it would be necessary to select either 000 or 001. While the 802.1p bits can only be one of four values, it is not necessary to rewrite the bits in the higher layer protocols that will be maintained across the network.

9.4.1   DIFFSERV INTEGRATION WITH MPLS

In order to integrate DiffServ with MPLS, two types of traffic engineered LSPs have been developed. The first type uses the EXP bits to carry the class of service information and is known as an E-LSP. Since DiffServ uses 6 bits and there are only 3 EXP bits, it is clear that each of the EXP bits has to represent an average of 8 DSCPs. It is only necessary to create a single LSP between the ingress and egress LSRs and simply identify the class of service by the EXP bits. Since it is possible to simply configure a mapping from EXP to Diffserv and back again on every LSR, there is no signalling required at the time the LSP is established. This means that the amount of MPLS state required is kept to a minimum. However, this reduction in state is at the cost of a reduction in granularity.

The second type uses a different label to represent each of the different classes of service. They are known as L-LSPs. One LSP is created between the ingress and egress LSRs for each class of service and traffic for each class is associated with the relevant FEC. In order to achieve this, it is necessary to signal the class of service at the time the LSP is established. This requires an additional RSVP object (DIFFSERV) or an additional LDP TLV (DiffServ). This makes it possible to exactly map each DSCP to a unique LSP. However, this increased granularity comes at the cost of significantly increased MPLS state.

10

Multicast

Multicast has the potential to vastly improve the scalability of networks carrying traffic matching particular profiles and types. It has significant potential to reduce traffic volumes when a relatively low number of sources transmit the same content to a relatively large number of receivers. If this model were supported with pure unicast, then the source would have to sustain the transmission of very large numbers of flows and very large aggregate volumes of data. With multicast, a single stream of data is transmitted by the source and is passed along towards all interested receivers, being replicated only when two receivers can be reached only via different paths.

Multicasting IP data requires a number of protocols, each of which performs a particular portion of the required functionality. Multicast has developed in much the same way as unicast, with separate protocols for communication between host and router and between router and router. However, multicast has the added feature of requiring routing protocols from source to destination and from destination to source!

Multicast traffic is sent by one or more sources (S) to a group (G). Interested hosts join groups. Multicast groups fall into the address range 224.0.0.0/4.[1]

10.1 MULTICAST FORWARDING AT LAYER 2

Most of this chapter focuses on multicast forwarding at Layer 3. However, it is equally important to be able to identify traffic on a multi-access network destined for a particular multicast group at Layer 2.

---

[1] For a complete list of allocated multicast addresses, see http://www.iana.org/assignments/multicast-addresses.

Designing and Developing Scalable IP Networks   G. Davies
© 2004 Guy Davies   ISBN: 0-470-86739-6

10.1.1   MULTICAST ON ETHERNET AND FDDI

It is necessary to provide a mapping from the Layer 3 multicast group into a Layer 2 multicast address. In an ideal world, this would be a one-to-one mapping. However, rather than mapping the lowest-order 28 bits from the IP multicast group address, the lowest-order 23 bits from the IP multicast address are appended directly onto a MAC prefix. This leaves five bits of the IP multicast group address that can vary while still resulting in an identical Layer 2 multicast group address. Thus, each Layer 2 multicast address maps to 32 Layer 3 multicast groups. The MAC prefix is 0x01005E in the first 24 bits of the MAC address plus 0 in the 25th bit of the MAC address. In order to have a one-to-one mapping, it would have been necessary to obtain 16 OIDs from IEEE. Because each of these OIDs cost $1000, the researchers who were first working on the development of multicast found it impossible to justify the purchase. Therefore, one OID was purchased by the researchers and shared with another project.

10.1.2   MULTICAST OVER TOKEN RING

Token Ring uses functional MAC addresses for devices performing special functions on the ring. The first bit of the first byte of the Token Ring MAC is used to signify whether the destination is unicast or multicast/broadcast. This bit is known as the I/G bit. The second bit of the first byte is used to signify whether the MAC address is a globally significant manufacturer assigned address or a locally significant address assigned by the local administrator. This bit is known as the U/L bit. The first bit of the third byte is used to signify whether an address is a functional address or a locally administered address. This bit is called the FAI bit.

In order to create a functional multicast address, exactly one of the 31 bits following the FAI bit is set to 1. Therefore, it is impossible to directly map IP multicast groups onto multicast Token Ring MACs. It is necessary to employ other means to overcome this constraint. Two mechanisms are currently employed.

The first mechanism uses the broadcast address FFFF.FFFF.FFFF for all multicast frames in addition to true broadcast frames. This is somewhat inefficient because all hosts receive the IP multicast frames, whether or not the hosts are interested in IP.

The second mechanism uses a single functional address C000.0004.0000 for all traffic destined for an IP multicast group. At first sight, this mechanism is a better option, but it relies upon the ability of every NIC on the ring to support and correctly handle this functional address, which many do not. In addition, even if the NIC on the host supports this functional address, the arrival of each multicast packet triggers an interrupt to be sent to the host's CPU. This can significantly impact the performance of any host on a ring that handles anything more than the smallest amount of multicast traffic. For these reasons, running multicast over Token Ring scales poorly.

10.1.3   INTERNET GROUP MANAGEMENT PROTOCOL (IGMP)

IGMP is the protocol used by a host to inform the nearest multicast-capable router of the multicast groups it wishes to receive. At the time of writing, there are three versions of IGMP, each of which provides a number of enhancements over the previous version.

IGMPv1 provides a simple mechanism for hosts to inform the nearest multicast router that they wish to receive a particular group. This message is known as a report. In addition, IGMPv1 provides a mechanism for routers to query whether any hosts on the LAN are interested in receiving multicast transmissions. IGMP queries are transmitted at regular intervals, the default being 60 seconds, onto the LAN. IGMPv1 provides no explicit mechanism for leaving a group, so even after the last interested host stops sending reports, the multicast router forwards traffic onto the network until the last report times out. If other hosts on the LAN are still interested, their reports ensure that traffic still makes its way onto the LAN. A host has 10 seconds (fixed in IGMPv1) to respond with a report to queries sent by multicast routers if they want to continue to receive any traffic. Because reports are sent to the group the host wishes to receive, all interested hosts receive the report sent in response to a query and do not send duplicate responses to queries.

In addition to the functionality of version 1, *IGMPv2* provides an explicit leave function. IGMPv2 also enhances the concept of the query by making it possible for a router to query for members of a specific group, in addition to being able to send a general query. These are major enhancements over version 1 and significantly improve the speed with which multicast traffic for one group can be stopped and, by implication, traffic for another group started. Consider, for example host H that is currently receiving traffic for group G1. H now wishes to stop receiving traffic for group G1 and start receiving traffic for group G2, instead. If it is using IGMPv1, then H will just stop transmitting reports for G1 and start transmitting reports for G2. Until the membership of G1 times out, H will receive traffic for both G1 and G2. If it is using IGMPv2, then H will send an explicit leave for G1 and a report for G2. The leave triggers the router to send an IGMP query (could be a group-specific query or a general query). Assuming no other hosts on the LAN are interested in G1, then the router will stop transmitting traffic for that group onto the LAN.

IGMPv3 provides the means for the host to specify the source from which it wishes to accept data for a particular group (i.e. rather than joining a group with an unspecified source (*, G), the host sends joins, which are analogous to reports in IGMP versions 1 and 2, for a group with a specified, single source (S, G)). This is specifically required for source-specific multicast (see below).

At the time of writing, IGMPv1 and IGMPv2 are widely implemented in both hosts and routers. IGMPv3 is less widely implemented, particularly on host operating systems, although it is becoming more widely implemented. If IGMPv2 is available, it is unquestionably a better, more scalable option that IGMPv1. It is questionable whether IGMPv3 provides any significant benefit over IGMPv2 unless you are implementing SSM, in which case, you must use IGMPv3.

All versions of IGMP are constrained to a single hop at Layer 3 between hosts and the first multicast router. However, this does not totally protect IGMP from scalability issues. For example, consider the situation where there are many hosts in a large switched network. Each of the receivers is interested in a small subset of the total set of groups available. However, because the receivers are all effectively attached to a single router and they are all sending IGMP reports/joins for their required groups, the router will transmit the multicast traffic for all the groups onto the LAN. Because the switch is not aware of who is interested in which groups, it must transmit all traffic for all groups to all switch ports. Even if no single group accounts for more than a few tens or hundreds of kilobits per second of data, the aggregate of hundreds or even thousands of groups could amount to many megabits per second of data, which must be transmitted down every switch port. Thus, there is potential for entirely unwanted traffic to totally over-subscribe a port.

10.1.4　IGMP SNOOPING

IGMP snooping is one widely implemented mechanism for the constraint of the flow of multicast traffic through a switched network. It works by snooping IGMP reports being sent by hosts on each port and explicitly permitting traffic for the requested group to be transmitted onto the port over which the IGMP reports were received. Traffic for that multicast group is not transmitted on any port for which there has been no matching IGMP report received.

This, however, raises another problem, which relates to the ports to which the multicast routers are attached. They never send any IGMP reports (although the IGMP querier, or PIM-designated router, will send IGMP queries) and, therefore, will never receive packets for any multicast groups because the multicast traffic is blocked from the ports to which they are attached. It is possible to manually disable IGMP snooping on a per-VLAN basis, thus explicitly permitting all multicast traffic to those VLANs. However, doing so overrides the behaviour for all ports on the VLAN. Ideally, it would be better to specify the ports connected to the multicast routers and have them accept all groups while still snooping other ports in the VLAN. However, the need to configure ports is really an unnecessary administrative load (although many would consider it desirable for security reasons).

10.1.5　PIM/DVMRP SNOOPING

In order to overcome the need for manual configuration of those ports on which PIM or DVMRP routers are attached, an automated mechanism to overcome this is provided. PIM/DVMRP (see below for descriptions of PIM and DVMRP) snooping can be configured, in addition to IGMP snooping on all ports in a VLAN, so that the switch also snoops PIM or DVMRP packets and IGMP queries. Upon seeing these packets, it marks that port open for all multicast groups. This means that should a new router be installed on the VLAN, it is not necessary to manually configure the directly connected switch.

10.1.6   IMMEDIATE LEAVE PROCESSING

When a multicast router receives a leave from an IGMP neighbour, it does not immediately process that leave but, instead, waits for further reports to arrive for the same group to override the leave. However, on the switch port on which the leave was received, it is desirable to immediately stop the multicast traffic. Therefore, it is worth configuring immediate leave processing on the switch, if available, on ports connected to hosts. However, this behaviour can be exceptionally disruptive if it is configured on switch ports connected to multicast routers or on trunk ports connected to other switches.

10.1.7   CISCO GROUP MANAGEMENT PROTOCOL (CGMP)

CGMP is a Cisco proprietary protocol that operates between routers and switches. It provides similar functionality to that of IGMP without requiring Layer 2 devices to snoop Layer 3 information. However, in addition to constraining the distribution of multicast traffic to only those ports with interested hosts, it also creates an interaction between switches and routers so that switch ports connected to routers can be identified and all constraints on multicast traffic is removed from those ports. Remember how a special feature had to be added to IGMP snooping to achieve this. Although both routers and switches must be configured for CGMP and to understand CGMP packets, CGMP packets are generated and sent only by the routers.

CGMP joins are sent to inform switches to add members to a multicast group. CGMP leaves are sent to inform switches to remove members from a multicast group or to delete the entire group. Both these messages have the same format and are sent to the same MAC address, 0100.0cdd.dddd. Both types of packets contain one or more pairs of MAC addresses, the Group Destination Address (GDA) and the Unicast Source Address (USA).

On starting up, a router sends a CGMP join to attached switches. These initial CGMP joins contain two addresses: the GDA is set to 0000.0000.0000 and the USA is set to the router's own MAC address. This join message informs the switch that a router is attached to the port on which the message was received. Every 60 seconds, another similar message is sent as a keepalive.

As normal, a host wanting to join a multicast group sends an IGMP membership report. Also as normal, the switch keeps a note of the MAC address of the attached hosts and the port to which they are attached. On receipt of the IGMP membership report, the router sends a CGMP join back to the switch with the GDA set to the multicast MAC address of the group and the USA set to the MAC address of the host. The switch can now map the multicast MAC address to a set of ports, and when it receives packets from the router destined for the multicast MAC address, it can forward the packets only to those ports mapped to the multicast MAC address. The packets are not forwarded to the router from which they are received.

IGMP queries are sent to all ports with connected hosts known to be interested in any multicast group. IGMP membership reports are forwarded to all routers.

When a host sends an IGMPv2 leave, the messages are forwarded by the switch to the routers. The router then sends IGMP group-specific queries, which are forwarded to all group ports. If another host responds, then the router sends a CGMP leave packet back to the switch with the GDA set to the multicast group MAC and the USA set to the MAC of the leaving host. If no host responds, then it is assumed that no host is listening to that group and the router sends a CGMP leave packet back to the switch with the GDA set to the multicast group MAC and the USA set to 0000.0000.0000. On receipt of this message, the switch removes the multicast group entirely.

10.2 MULTICAST ROUTING

10.2.1 REVERSE PATH FORWARDING (RPF) CHECK

A fundamental element of multicast routing is the RPF check. Multicast routing protocols require a router to understand not only where the traffic for a particular group needs to go but also the interface through which the source can be reached via the shortest path. The RPF check is the process whereby unicast forwarding state, established from the router towards the root of the multicast tree, is used to identify the interface over which multicast traffic from that source should be received. The router checks that it is receiving data for a particular (S, G) on the interface through which it would reach the source (S). In addition, the RPF check includes a check that the next hop back towards the source is a neighbour in the multicast routing protocol.

10.2.2 DENSE MODE PROTOCOLS

Dense mode protocols are most suited to environments in which receiving hosts for groups are attached to a large proportion of the leaf routers. Dense mode protocols work on the basis of 'flood and prune' distribution. All traffic received by a multicast router operating in dense mode is transmitted on all ports enabled for multicast, except that over which the traffic was received. It is stopped only if an explicit 'prune' is received from a downstream neighbour. When a prune is received by an upstream router, that router sets an expire timer to a default value of 2 hours that counts down to zero. Once it reaches zero, the interface is put back into forwarding state and it starts transmitting again until it receives a prune again. If, having transmitted a prune, a leaf router receives an IGMP join, it can send an explicit join back upstream and the upstream router will start transmitting immediately. Without the explicit join, the host would have to wait for the prune to expire.

10.2.2.1 Distance Vector Multicast Routing Protocol (DVMRP)

DVMRP is a self-contained multicast routing protocol that creates its own RPF topology information using a derivative of RIP. It does not rely upon any external unicast routing protocol. This RPF topology information is potentially redundant in the sense that almost any network running a multicast routing protocol will also be running a unicast routing protocol from which RPF topology information can be derived. However, the positive aspect of this is that it is possible to create a divergent multicast topology without requiring modifications to the unicast routing protocols.

At the time of writing, there are three versions of DVMRP. Versions 1 and 2 identify neighbours upon receipt of route advertisements from other routers on the subnet. If multiple routers are detected on a subnet, the designated router, which is responsible for sending multicast frames and IGMP queries onto the subnet, is elected based on the router with the lowest IP address on the subnet.

DVMRP version 3 starts by establishing neighbour relationships by sending probe packets, which contain a variety of information including flags describing the capabilities of the originator, a generation ID to identify the change of state of a neighbour (more on this later), and a list of all neighbours on the same subnet from which the originator has already received probe packets. When a router receives a probe packet from a neighbour with its own address in the list of neighbours, it knows that a two-way relationship has been established. Once the session is established, both neighbours will continue to send probes at 10-second intervals as a keepalive. Neighbours are considered dead if no probe packet is received within 35 seconds. By changing the configured probe interval, it is possible to configure the dead time.

Now, back to the generation ID. When a router recovers from a failure, it re-establishes its neighbour relationships. As stated above, one of the elements of the probe packet is the generation ID. This is a consistently increasing 32-bit identifier arbitrarily derived locally by the originating router. When a router receives a probe from a 'known' neighbour with a different generation ID, it recognizes that the neighbour has restarted. Upon seeing this, the receiving router flushes all prune information received from that neighbour and unicasts a complete copy of its routing table to the restarting router. This allows a restarting router to restore its routing topology and commence forwarding multicast packets in the quickest possible time. Without this mechanism, the restarting router would have to wait for the prune state on the upstream routers to expire before it received any multicast packets and for the next periodic routing update to repopulate its routing table.

DVMRP version 3 also differs from versions 1 and 2 in the mechanism used to elect a designated router. It simply uses the IGMPv2 querier rather than running a separate election.

As mentioned above, DVMRP maintains its own independent routing topology using a derivative of RIP. Each router sends report messages to its neighbours containing the complete routing table plus all directly connected multicast-enabled subnets. These reports are sent every 60 seconds to the multicast group 224.0.0.4 (all DVMRP routers). If a known route is not refreshed for 140 seconds, that route is considered to be inactive, at which point the route is placed in hold down for a further two report intervals (120 seconds), during which time it is announced with a metric of infinity (32). If the route still has not been refreshed at the end of the hold down timer, it is flushed from the routing table. The topology created by this exchange is used to ensure that multicast packets are accepted only on the interface through which a unicast packet would be sent back to the source.

10.2.2.1.1 Downstream Dependencies

In addition to the RPF topology, upstream routers are informed of the dependencies placed upon them by downstream routers. This is achieved by the downstream routers sending a 'poisoned reverse' route back to the upstream router from which they learned the route. The poisoned reverse route is created by sending the same prefix with its metric set to the received metric plus infinity (32). Thus, a prefix received with a metric of 7 is returned upstream with a metric of 39. On receipt of this announcement, the upstream router recognizes that the downstream router is relying upon it for this RPF entry. This means that if potentially contradictory requirements are expressed by two or more downstream routers, upstream routers can ensure that the decisions they make with respect to forwarding or blocking on particular interfaces are appropriate to ensure that all those requiring the multicast traffic receive it.

10.2.2.1.2 DVMRP Designated Forwarder

As with all multicast topologies, it is possible for multiple copies of a multicast packet to be forwarded onto a single segment by several connected routers. In order to reduce the effects of this, DVMRP elects a designated forwarder on shared networks. If one of the routers is topologically closer to the source, it is selected as the designated forwarder. If two or more routers are topologically equidistant from the source, the router with the lower IP address is selected as the designated forwarder.

10.2.2.1.3 Forwarding Multicast Packets

When a DVMRP router receives the first packets to a new (S, G), it performs an RPF check to ensure that the packets were received on the upstream interface. If not, the packets are silently dropped. If so, a new (S, G) entry is created in the forwarding table and the packets are forwarded to all downstream routers recorded as being dependent upon that router. If the router is the IGMP querier for a particular shared network, it also sends an IGMP query onto that network. If the router receiving the multicast packets has no downstream routers and receives no responses to IGMP queries, then it sends a prune message upstream.

DVMRP prunes contain a prune lifetime, which informs the upstream router how long it should retain the state before again flooding the multicast traffic. By default, the prune lifetime is 2 hours.

If, having pruned itself from the multicast tree for a particular (S, G), a router receives an IGMP join for that (S, G), it must send a graft message upstream to reconnect itself to the multicast tree. Grafts are sent back upstream, hop by hop, until a router on the active tree is reached. Grafts are acknowledged at each hop with a graft ack message. This ensures that if no traffic is received following a graft being sent upstream, the router can differentiate between a lost graft message and a nontransmitting source.

10.2.2.1.4 DVMRP Implementations

DVMRP is relatively widely implemented by router vendors. However, some vendors have implemented only those elements of the protocol required to provide basic interoperability and the ability to integrate with routers that do not support PIM (see below).

Where possible, most service providers use PIM rather than DVMRP because of the perceived redundancy of much of the information maintained by DVMRP and the generally incomplete implementations of DVMRP from several vendors. Also, DVMRP suffers from the slow convergence associated with RIP. As you will see below, PIM defaults are not much better but most implementations provide the facilities to make PIM converge more quickly.

10.2.2.2 Protocol Independent Multicast — Dense Mode (PIM-DM)

PIM-DM works in a very similar way to DVMRP. However, it does not have any means to create its own RPF topology, nor does it have any mechanism to inform upstream routers of the dependencies placed upon them by downstream routers (more on this later). PIM (in sparse or dense mode) is independent of the underlying unicast protocol from which it derives RPF information. This is the origin of the 'protocol independent' part of its name.

PIM-DM floods multicast traffic on all enabled interfaces until it receives a prune on that interface. On receipt of a prune, a timer is set, by default to 210 seconds, and starts counting down towards zero. When the timer reaches zero, multicast traffic is flooded again to all enabled interfaces. If no prune is ever received on an interface, a router transmits multicast traffic indefinitely or until the source stops transmitting.

10.2.2.2.1   PIM Forwarder

If two or more PIM routers are attached to a single multi-access network, then it is inefficient if all of them transmit a copy of the same packet onto the network. Therefore, it is necessary to select one of them to forward packets. The selected router, not surprisingly, is known as the PIM forwarder. The election process is relatively simple. It is triggered when a PIM router receives multicast traffic from a neighbour on an interface on its outbound interface list for the particular (S, G). On detecting this situation, each router sends a PIM assert message. Among other things, this message contains the address of the interface connected to the multi-access network, the metric preference, and the metric of the route back to the source. The selection of the PIM forwarder is based on the following process:

1. The router with the lowest metric preference (administrative distance in Cisco terminology) associated with the route back to the source is preferred. This occurs only if the routers learn the route back to the source from different unicast routing protocols.

2. If all routers have routes with equal metric preference, then the router with the lowest metric associated with the route back to the source is preferred.

3. If all routers have routes with equal metrics, then the router with the highest IP address on its interface attached to the multi-access network is preferred.

The losing routers prune their interfaces from the multi-access network and set the expire timer so that only the winner transmits traffic for the specific (S, G) onto that network.

This can cause significant delays for traffic to recover after the winner's connection to that multi-access network fails.

10.2.2.2.2   PIM Prune Override

If there are two downstream routers on a multi-access network, only one of which has downstream receivers, a situation arises in which the router with no downstream receivers sends a prune. However, this prune has the potential effect of pruning the entire multi-access network, which is patently not what is intended. You will recall from the previous section that DVMRP propagates information regarding the dependencies placed on upstream routers by downstream routers. However, no such information is propagated by PIM-DM. Therefore, the prune sent by the router with no downstream receivers must be explicitly overridden by any router that actually has interested downstream receivers. This is achieved by sending a join back onto the multi-access network for the relevant (S, G). This overrides the prune sent by the router with no receivers.

10.2.2.2.3   Resilience and Scaling with PIM-DM

As mentioned briefly above, there can be significant issues when trying to create resilient networks using PIM-DM. This is particularly true if extremely rapid recovery is of very great importance. There are several timers associated with PIM, each of which can introduce delays in the restoration of a valid multicast tree from source to all receivers. The obvious candidates are the assert timer, the prune expire timer, and the PIM hello timer.

The prune expire timer is, by default, 210 seconds. Normally, this can be overridden by sending an explicit join or graft upstream to reconnect a link to the multicast tree without having to wait for the timer on the upstream interface to expire. However, if a prune is initiated by the loss of a PIM forwarder election, reconnecting the link to the multicast tree can be difficult and in some cases impossible until the assert timer expires. Consider the network in Figure 10.1. R3 and R4 elect a PIM forwarder on LAN C. For this example, we can assume that R3 becomes the PIM forwarder, which causes R4 to prune its interface from LAN C. If R3's interface on LAN C is, for whatever reason, disconnected from LAN C, there is no way for R4 to be reconnected before the assert expire timer on R4's interface times out. There is no downstream router to send a join back to R4, so no downstream hosts will receive any packets for that specific group until the assert expire timer reaches 0.

If we look at LAN B, R1 and R2 also have to run a PIM forwarder election. For this example, assume that R1 wins the election. R2 prunes its interface onto LAN B. If R1's link to LAN B is broken subsequent to winning the election, then R1, R3, and R4 will notice that their PIM neighbour has gone down only after 105 seconds. The loss of the PIM neighbour would cause R3 and R4 to lose their RPF next hop, which in turn causes an RPF check to fail. As the unicast routing protocol converges, taking up to 180 seconds if you're using RIP, the RPF next hop for that (S, G) is updated. At the point where the RPF check and routing protocol converge, a PIM prune would be sent to the old RPF next hop and a join would be sent back up to the new next hop to re-establish the flow of traffic.

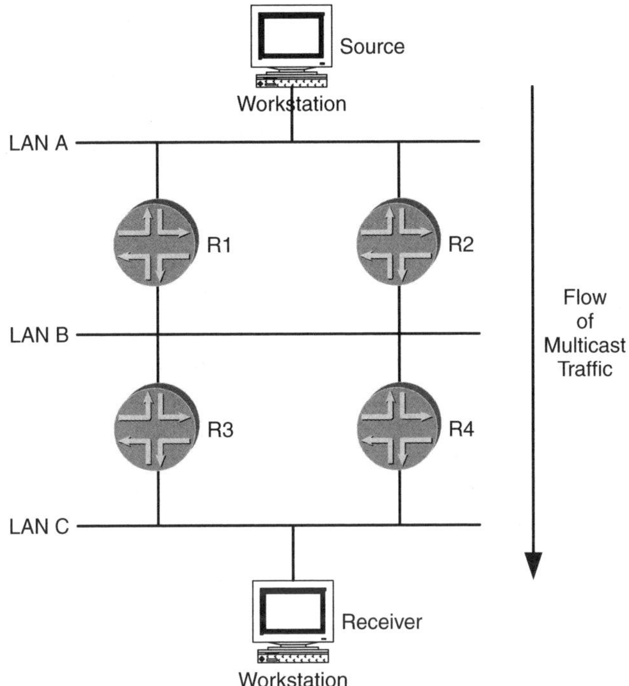

Figure 10.1 Example multicast network

10.2.3 *SPARSE MODE PROTOCOLS*

There are a number of sparse mode protocols but by far the most widely implemented is PIM sparse mode (PIM-SM). Most of the issues that apply to PIM-SM also apply to the other sparse mode protocols. Therefore, I will describe just the basic operation of PIM-SM. This will be used as the basis for a discussion of the general issues affecting sparse mode protocols. Other protocols may use slightly different terminology, but the underlying operation is fundamentally similar.

Unlike dense mode protocols, sparse mode protocols are particularly efficient when receivers are sparsely distributed across a network with a large number of leaf routers. Assuming that the leaf router is not already receiving multicast traffic for a particular group when it receives an IGMP membership report for that group from an attached host, it must send a graft towards a rendezvous point (RP). The RP is a central device that creates a shortest path tree back to all known sources in the domain and is responsible for forwarding the received traffic to hosts that join a shared tree or RP tree (RPT) with the RP as its root. At each hop along the way, the router checks whether it is already on the shared tree for that group and, if so, grafts the interface over which the graft was received onto the shared tree for that group. If not, it sends a graft upstream towards the RP, and the same process is repeated. The graft causes a new branch to be created on the RPT. Once traffic is received via the RPT, the leaf router can derive the location of the source of the group from the packets arriving destined for the group. Some sparse mode protocols allow for the creation of a branch to a shortest path tree (SPT) connecting the leaf node and the router adjacent to the source of the multicast stream by the most direct route. The main reason for this is to reduce the load on the RP. If all leaf nodes stayed on the RPT rather than joining the SPT, then the load upon the RP could become unsustainable. By joining the SPT for a particular group, the leaf node potentially causes its branch of the RPT for that group to be pruned.

10.2.3.1 Protocol Independent Multicast — Sparse Mode (PIM-SM)

PIM-SM is by far the most widespread multicast routing protocol in use by service providers today. It is particularly well suited to the sparsely located receivers you are likely to find in service provider networks.

Despite the name, in many respects PIM-SM has less in common with PIM-DM than PIM-DM does with DVMRP. PIM-SM requires the creation of an RP. The RP, as the name suggests, is the router that provides a rendezvous point for shortest path tree created from the designated router (DR) adjacent to the source and the RP tree from the DR/PIM forwarder adjacent to the receiver.

When a new source starts transmitting, the PIM DR on the network to which the source is attached starts sending unicast register messages to the RP. This triggers the RP to create a SPT back towards the source. While the SPT is being created, all multicast traffic is sent to the RP encapsulated within register messages. This immediately places constraints upon any device that might be the DR or the RP. Each has to encapsulate and de-encapsulate packets for brief periods but potentially at a relatively high speed. Some vendors require special hardware to support the function of DR or RP, while others simply pass all packets requiring encapsulation or de-encapsulation to the central processor.

In order to mitigate the potentially severe impact this can have on the CPU of the router, Cisco provides the ip pim register-rate-limit command to reduce the number of register messages that can be sent by a DR. This reduces the potential load upon the DR and the RP at the expense of possible packet loss if more than the specified number of packets is being sourced during the register process.

All leaf nodes to which interested hosts are attached join the RPT through which traffic is delivered. After a configurable period, the leaf node can create an SPT directly to the source. This helps to reduce the load on the RP and removes less than optimal paths.

10.2.3.1.1 Identifying the RP

As mentioned above, PIM-SM requires an RP. The simplest mechanism for identifying the RP is to statically configure it on every router in the multicast domain. However, this can lead to major problems with scalability and, more specifically, with resilience. If every router in a domain had a single, manually configured RP and that RP were to fail, then it would be impossible for any router to join the RPT to receive new streams. In addition, any host receiving streams via the RP would lose the stream and be unable to re-establish connectivity until the RP recovered.

It is possible to specify which groups are served by a particular RP. Thus, manually sharing the load of groups between a number of RPs is a relatively simple, if tedious matter of configuring different RPs and listing ranges of groups serviced by each. However, if one of the RPs fails, no other RP will be waiting in the wings to take over the job. Traffic for the groups served by the failed RP will not be delivered to any hosts not already attached to a SPT.

Three methods have emerged to provide automated resilience of the RP function. One (auto-RP) was developed by Cisco and is effectively proprietary, although a number of other vendors have implemented auto-RP. The second (the bootstrap protocol) was developed by the IETF and has been implemented by most, if not all, vendors. Finally, there is the anycast-RP. This is not a new protocol or enhancement to an existing protocol. Instead, it relies on an inherent behaviour of existing unicast routing protocols to find the nearest RP from a set of RPs that all use the same address.

10.2.3.1.1.1 Auto-RP

Auto-RP is a mechanism for routers throughout a sparse mode multicast domain to automatically discover the RP associated with each multicast group. Cisco developed auto-RP before the bootstrap protocol was specified in the IETF PIM working group. Auto-RP uses two well-known reserved groups, 224.0.1.39 and 224.0.1.40, to exchange information between candidate RPs, mapping agents and listeners.

Each candidate RP is configured with a set of groups for which it provides service. Every 60 seconds, the candidate RP announces its own IP address and a list of the groups for which it is a candidate RP to the multicast group 224.0.1.39. Mapping agents are configured to listen for the messages transmitted by the candidate RPs and select the candidate RP with the highest IP address for each group. The mapping agent then announces all of the selected group-to-RP mappings to the group 224.0.1.40. All other routers in the domain are configured to discover the RPs by joining the group 224.0.1.40.

Auto-RP presents the operator with an interesting problem. The RP-announce and RP-mapping messages are sent to multicast groups and need to reach all multicast routers throughout the domain. The problem is that auto-RP is inherently associated with a sparse mode domain. Therefore, in order for the routers to join the groups you need an RP, but you will not be able to identify the RP until the RP-mapping messages arrive. So, you have a Catch-22 situation. There are two possible solutions to this problem. The first solution is simply to statically configure every router in the domain with the RP for the two groups, 224.0.1.39 and 224.0.1.40. This is simple in the sense that it does not require any special protocol changes. However, it lacks the resilience, which is the whole point of using auto-RP. The second option is to use a hybrid mode in PIM called PIM sparse-dense mode. This hybrid mode permits the operator to declare specific groups that are to be carried in dense mode. All other groups are carried in sparse mode. So, with sparse-dense mode, you can define 224.0.1.39 and 224.0.1.40 as dense groups throughout the domain. Thus, the RP-announce messages and RP-mapping messages can reach all parts of the network, and the RP can be selected and announced.

10.2.3.1.1.2 *The Bootstrap Protocol*

The bootstrap protocol was defined by the PIM working group of the IETF. It is technically still an Internet draft and has not yet made it to full RFC status. The bootstrap protocol defines two types of devices. Candidate bootstrap routers (C-BSR) and Candidate RPs (C-RP) are manually configured within the domain. Each C-BSR announces its address to the ALL-PIM-ROUTERS multicast group on 224.0.0.13.

The first step in the bootstrap process is to elect a bootstrap router from the C-BSRs. Each C-BSR is assigned a priority from 0 to 255, 0 being the default. When a C-BSR first starts up, it sets a bootstrap timer to two times the bootstrap period plus 10 seconds (defaults to 130 seconds) and listens for messages from a BSR. If a message is received from a BSR with a lower priority than the local C-BSR, the C-BSR resets the bootstrap timer and listens for another message. If a message is received from a BSR with a higher priority than the local C-BSR, the C-BSR declares itself BSR and sends bootstrap messages every 60 seconds (the bootstrap period). If the C-BSRs have the same priority, the C-BSR declares itself BSR if it has the higher IP address. If no message is received by the C-BSR by the time the bootstrap timer expires, it assumes that no BSR exists, declares itself BSR, and starts transmitting bootstrap messages.

Bootstrap messages are always transmitted with a time-to-live (TTL) value of 1 to the all PIM routers group. On receipt of a bootstrap message, each PIM router transmits the message on all PIM-enabled interfaces except the one on which the message was originally received. This 'dense' behaviour ensures that the messages are propagated throughout the multicast domain without the need for the special sparse-dense behaviour required for auto-RP.

Each C-RP is configured with an RP address, a priority between 0 and 255, and a list of groups for which it is an RP. When the C-RPs receive the bootstrap messages, they start transmitting candidate RP advertisement messages directly to the BSR. These messages contain the parameters listed above.

The BSR receives the candidate RP advertisements from the C-RPs and compiles an RP set from all the advertisements. The RP set is then included in the bootstrap messages that are being propagated throughout the domain. In addition to the RP set, an 8-bit hash mask is included in the bootstrap message.

At the point where a leaf node is triggered to join an RPT, it examines the RP set and the hash mask and selects the RP for the particular group based on the following criteria:

1. If there is only one C-RP for a group, that is the RP for that group.

2. If there are two or more C-RPs for a group, the priorities of the C-RPs are compared. The C-RP with the lowest priority is selected RP for that group.

3. If the lowest priority is associated with two or more of the C-RPs for a group, a hash function is run. The input to the function is the group address, the hash mask and the RP address. The output is a single numeric value. The C-RP with the highest output value is selected RP for that group.

4. If the highest output value is output from the hash function for two or more of the C-RPs in the election above, the C-RP with the highest IP address is selected RP for that group.

This mechanism ensures that every router in the multicast domain will select the same RP for the same group. The hash function is used to ensure that an arbitrary number of consecutive groups are served by the same RP.

10.2.3.1.1.3 Anycast RP

Anycast is really not a specific feature of PIM. However, it is widely used as a mechanism for creating a set of RPs from which a router selects the nearest. This process is entirely independent of PIM and is really a feature of the underlying IGP. Each RP is configured with the same local address, usually on a loopback interface that is announced via the IGP. Every router in the network is then configured with the single RP address, the anycast address. The RP, which is used, is based entirely upon the IGP metrics to each of the routers announcing the anycast address.

However, this mechanism raises some problems of its own. The first problem is that if left as described above, each RP will become responsible for a subset of groups and sources in the network and that subset of groups will only be accessible to a subset of hosts in the network. The set of hosts using a particular RP as sources and as receivers is identical.

In order to solve this problem, it is necessary for RPs to tell each other which active groups they are supporting. This is achieved with MSDP, which is described in more detail later in this chapter. MSDP was originally designed to support interdomain multicast routing. However, in this case, it performs an identical function within a single domain.

Once each of the RPs knows the total set of (S, G) that is supported by the complete set of RPs within the network, each RP can join the SPT towards every source, whether local or served by a different RP. This allows any client to receive traffic for any (S, G) within the entire network. Because all RPs know all the (S, G) on all RPs in the domain, then if one RP fails, the IGP will converge and all clients previously using the failed RP will now find that a different RP is closest. The outage as seen by the client would be the time taken for the IGP to converge, for the DR to join the RPT to the new closest RP and for the RP to join the SPT back to the source (it is possible that the RP would already be on the SPT to the source, thus the outage would be further reduced).

10.2.3.2 PIM Designated Router

On multi-access networks to which multiple PIM routers are attached, one router must take responsibility for sending the register messages to the RP to trigger the creation of the SPT for traffic from sources on that multi-access network. The DR is responsible for encapsulating and de-encapsulating the traffic to the RP until the SPT is established. The DR is also responsible for performing the IGMP queries on IGMPv1 LANs, because IGMPv1 has no mechanism for electing a querier of its own.

10.2.3.3 PIM Forwarder

PIM-SM uses an identical process for electing a PIM forwarder. The forwarder performs exactly the same function in PIM-SM as it does in PIM-DM.

10.2.3.4 PIM Prune Override

PIM-SM uses exactly the same process as PIM-DM for one downstream router on a multi-access network to override prunes sent by another downstream router on the same multi-access network.

10.2.3.5 Source-Specific Multicast (SSM)

SSM is a further development of multicast using PIM-SM. It changes the basis of multicast from the situation where an arbitrary number of sources will transmit to a group and a receiver will take multicast traffic from any source to the situation where the receiver specifies the source from which they will accept data for a particular group. A single (S, G) is known in SSM terminology as a channel.

In order for a host to indicate its desire to join a group with a specific source, that host and the attached PIM routers must support IGMPv3. When a source-specific join is received, the DR immediately creates a SPT directly to the source without joining a RPT first, so the need for a RP is removed.

SSM provides some added benefits with respect to scalability over and above those inherent with the behaviour of multicast. For example, the fact that a stream of data transmitted to one group from one source is completely distinct from a stream of data transmitted to the same group from a different source means that the problems associated with global group allocation are totally alleviated. Now, group addresses have only to be unique with respect to a single source. Because the range 232.0.0.0/8 has been allocated by IANA for use with SSM, it effectively means that each source could transmit streams to 2^{24} groups!

SSM also improves the control of content on a particular channel. Because only one source can transmit onto one channel, it is possible to effectively filter any possible spurious data being injected into a channel.

Finally, the removal of the requirement for an RP means protocol extensions like auto-RP and the bootstrap protocol become unnecessary. Also, if there are no RPs, then there is no need for MSDP. Interdomain multicast is simply a case of connecting your PIM-SM domain to that of your neighbours.

10.2.4   MULTICAST SOURCE DISCOVERY PROTOCOL (MSDP)

One major scaling issue with multicast is how to exchange multicast group information between multicast domains. One way would be to share RPs, but this has severe limitations with respect to scaling and also requires a significant amount of trust on the part of those who have no direct control over the operation of the RP. This is not a realistic solution for service providers in today's market. Therefore, an alternative is required. The alternative protocol, which was originally designed as a stopgap until a proper multicast BGP became available, is MSDP.

MSDP exchanges information about active multicast groups and the sources of those groups between independent RPs in one or more multicast domains. Each MSDP router peers directly with other RPs. The RPs send each other messages containing details of groups and the associated active sources.

In order for a host in one domain to join a group that has a source in another domain, the host sends the normal IGMP message, which is translated to a PIM-SM join to the local RP. The local RP then joins the SPT direct to the source in the other domain.

10.2.5 MULTIPROTOCOL BGP

As discussed. BGP provides the means to carry NLRI other than that for unicast IPv4 (for which it was originally designed). MBGP can carry multicast RPF information between ASs. MBGP does not carry group information. It is particularly useful when multicast information is being shared between ASs using MSDP between RPs.

10.2.6 MULTICAST SCOPING

Multicast scoping is required to ensure that multicast traffic can be constrained within a particular boundary. This is sometimes desirable because some traffic distributed over multicast has a constrained validity or may be of a nature which makes global visibility undesirable (e.g. commercially sensitive material or material for which a fee must be paid to gain access).

Two main mechanisms exist for scoping multicast:

1. TTL;

2. Administrative scoping.

Setting the TTL is a crude mechanism that can be used to create administrative boundaries by ensuring that packets pass no further than a certain number of hops from the source. It is notoriously difficult to use this mechanism to provide accurate, reliable, and consistent scoping of multicast packets. For example, the behaviour in the stable state might be dramatically changed in the presence of failures.

An example of the limitations of scoping based on TTL can be seen in this example. When using PIM-DM, scoping using TTL has some potentially severe side effects on the equipment at the administrative boundary. While in the lab, I came across this problem and spent quite some time working through the variety of symptoms. The network was as shown in Figure 10.2.

The video sources at the bottom of Figure 10.2 were transmitting multicast streams at 20 Mbps with the TTL set to 4. As you can see, packets sent from a CODEC attached to R3 reach the R4 with TTL $= 1$ so R4 would decrement the TTL and it would expire. On some routers, all packets of this type must be sent to the CPU for processing rather than

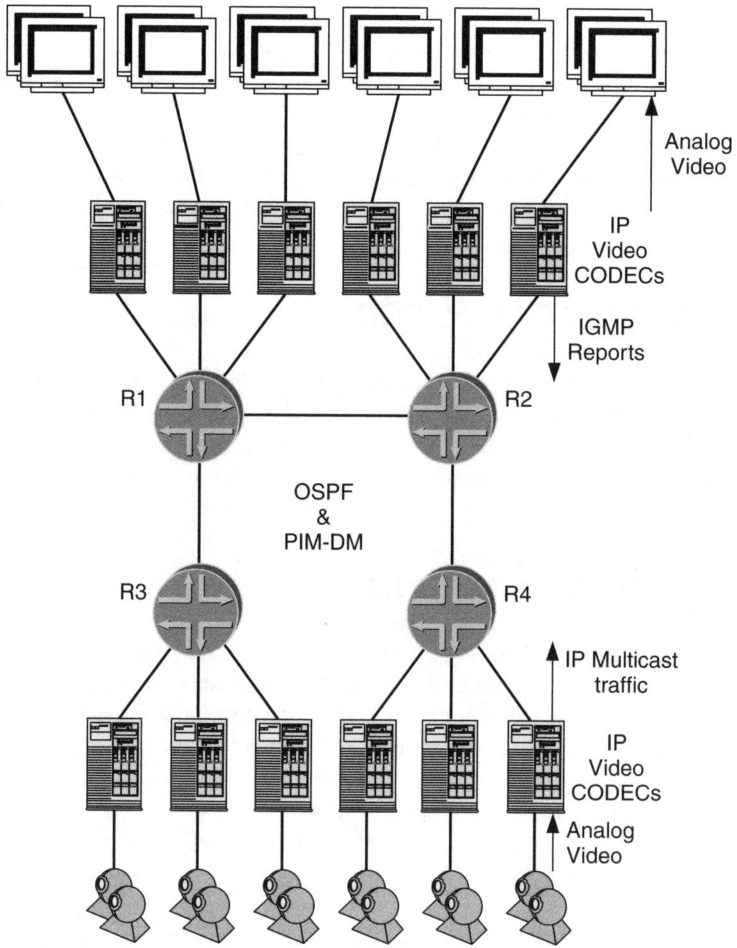

Figure 10.2 Example of the limitations of TTL-based multicast scoping

being switched in hardware. So, we had a stream of data arriving at 20 Mbps with all packets having an expired TTL. This sent the CPU of the devices to between 25 and 40% with just one video source. Worse, because the packets were arriving with TTL = 1 and were expiring on the router, they were not being processed by PIM so no prune was ever sent to R2 and the stream just kept on flowing. We did not actually spot the increased CPU until we had added many streams, all with a TTL of 4. The effect of all those streams, which were never pruned, was that the CPU of the DR became totally overwhelmed by the processing load of responding to all the TTL expired packets. The IGP neighbours and PIM neighbours could no longer be maintained and all the adjacencies were lost. This caused the CPU load to reduce, and as the IGP and PIM adjacencies recovered, the flows started again and the CPU load started to rise again. This cycle continued until the source of multicast packets was removed.

This issue could not be attributed to any single element in the network. The TTL of 4 on the IP video CODECs was not actually in place for the purpose of administrative scoping but was the default, the architecture of the routers R3 and R4 was such that they were particularly vulnerable to this problem and the fact that the packets were (correctly) not being processed by PIM exacerbated the problem by failing to prune the link so the flows never stopped. The fact that the flows were each running at 20 Mbps also exacerbated the problem because this equates to roughly 82,000 packets per second (256-byte packets), each of which must be sent to the CPU and generate an ICMP TTL Expired response.

In order to overcome the limitations of the TTL-based mechanism described above, it is possible to administratively scope a multicast network. A particular range of groups has been specifically defined for this purpose. The range is 239.0.0.0 to 239.255.255.255. Within that range, two subranges have been defined:

1. 239.255.0.0 to 239.255.255.255 is reserved for IPv4 site local scope. The definition of 'site local' in this sense is actually somewhat nebulous but traffic to these groups may not cross any administrative boundary of any kind.

2. 239.192.0.0 to 239.251.255.255 is reserved for IPv4 organization local scope. This range is available for organizations to allocate groups for private use. Routing information and traffic for these groups must never cross the boundary of an autonomous system.

The rest of the range, from 239.0.0.0 to 239.191.255.255 and from 239.252.0.0 to 239.254.255.255, is currently not allocated for any particular purpose but is reserved for expansion of the above two classes, each expanding downwards. These ranges should not be used until the existing scoped addresses have been exhausted.

In addition to 239.0.0.0/8, the range 224.0.0.0/24 has been allocated for well-known link local groups and 224.0.1.0/24 for well-known global scope groups. A number of other groups have been allocated specifically as global scope well known groups.

In addition to the definition of administratively scoped groups, it is necessary to define the boundaries of your network. This is achieved simply by identifying a port and the groups, which may not be forwarded via that port in either direction. That port represents the boundary.

11

IPv6

Since the late 1990s, it has been patently clear that the constraints of IPv4 were too great to support the ballooning demand for Internetworked services. Initially, the main focus of the limitations was on the address space with predictions of the exhaustion of IPv4 addresses ranging from early the next century (2000) through to some time over the next couple of decades. One suggested solution was that the limited address space could be overcome by using network address translation (NAT) in conjunction with private addresses as defined in RFC 1918. While NAT does provide the means for many hosts to appear as a single host (or as a small number of hosts), it introduces its own set of new constraints. Some of these new constraints are potentially more difficult to overcome or impose greater difficulties than those being solved. In particular, certain protocols do not behave at all well in the presence of NAT. However, in some cases, extensions have been added to the various protocols to try to mitigate some of those negative effects.

With this in mind, the IETF started a working group to design the next generation of the Internet Protocol in the IPng WG, which has subsequently been renamed the IPv6 Working Group. The base definition of IPv6 is RFC 2460.

For the curious among you, version 5 of the IP protocol was allocated to the Stream Protocol some time before the development of IPv6. It is a somewhat experimental protocol designed to work in conjunction with IPv4 and is based significantly upon IPv4 including using the same framing and addressing. One of its major perceived benefits was that it is a connection-oriented protocol, potentially providing improvements in the ability to provide guaranteed levels of service.

11.1 EVOLUTION AND REVOLUTION

IPv6 is both an evolutionary and revolutionary development of IPv4. As you would expect, all the base functionality that made IPv4 so successful has been reproduced, although

Designing and Developing Scalable IP Networks   G. Davies
© 2004 Guy Davies   ISBN: 0-470-86739-6

many of the protocols have been given different names, which can confuse the unwary (e.g. Address Resolution Protocol has been replaced by Neighbour Discovery). In many cases, these renamed protocols have also acquired extra functionality or have slightly modified behaviour. Other functions, previously in the realm of higher layer applications, have to a varying extent been subsumed into the base protocol (e.g. Stateless and Stateful autoconfiguration rather than DHCP).

11.2   IPv6 HEADERS

IPv6 dramatically changes the format of the header compared to that used in IPv4. Despite the extension of the address space from 32 to 128 bits (this obviously impacts the source address and destination address in the header), the basic IPv6 headers (40 bytes) are not proportionally larger than the IPv4 headers (24 bytes). This is achieved by creating many extension headers, none of which are mandatory. This provides a hugely flexible mechanism for efficiently adding extra functionality without increasing the overhead on the majority of packets.

11.3   IPv6 ADDRESSING

One of the first issues you will notice about IPv6 is the entirely unfamiliar addressing format. The IPv6 addressing architecture is defined in RFC 3513. Whereas IPv4 has 32-bit addresses presented as four decimal numbers separated by a dot with each decimal number representing 8 bits of address space, IPv6 has 128-bit addresses. These are represented as one to eight blocks of one to four hexadecimal digits. Each block of four digits is separated by a colon. This sounds slightly confusing but it is not so bad, really. Starting with the most basic example, the address is 32 hexadecimal digits in eight blocks of four:

1234:5678:9ABC:DEF0:0000:0000:0000:0001

Note that there are several blocks above in which some of the higher-order bits are zero. If there are leading zeroes in a block of four, they can be ignored so the above can become:

1234:5678:9ABC:DEF0:0000:0000:0000:1

Taking this a step further, if there is a 16-bit block that is entirely zero, the entire block can be reduced to two consecutive colons. This can be further compressed if there is more than one consecutive 16-bit block containing all zeroes. In this case, the entire string of zeroes can be compressed to two colons. Therefore, the address above can be reduced to

1234:5678:9ABC:DEF0::1

Note, the lower-order 0 in DEF0 cannot be removed. If it were removed, it would be equivalent to 0DEF, rather than DEF0.

The double-colon notation can only be used once in an address. If there were two sets of double-colons, it would be impossible to say how may 16-bit blocks were associated with each. For example, 1234::5678::9ABC could be equivalent to any of the following:

1234:0000:5678:0000:0000:0000:0000:9ABC
1234:0000:0000:5678:0000:0000:0000:9ABC
1234:0000:0000:0000:5678:0000:0000:9ABC
1234:0000:0000:0000:0000:5678:0000:9ABC
1234:0000:0000:0000:0000:0000:5678:0000:9ABC

The addresses above all lack one final thing. As with IPv4, IPv6 addresses can be conceptually broken into a part representing the network and a part representing the host. The number of bits representing the network is known as the prefix length. In IPv4, this is represented either as a dotted quad subnet mask or as a /prefix length. IPv6 has no concept of a 'dotted quad' or colon separated mask for that matter. It only uses the /prefix length notation. The prefix length is a number from 0 to 128. So the address above could be represented as:

1234:5678:9ABC:DEF0::1/64

In this example, 1234:5678:9ABC:DEF0 represents the network portion of the address and ::1 represents the host portion.

11.3.1   HIERARCHICAL ALLOCATIONS

Now you understand the way in which IPv6 addresses are represented, we should look at the way IPv6 addresses are hierarchically allocated to particular functions. The following are the main functional allocations, which have been created at the time of writing:

- 0200::/7—NSAP allocation
- 0400::/7—IPX allocation
- 2000::/3—Aggregatable Global Unicast Addresses
- FE80::/10—Link Local Addresses
- FEC0::/10—Site Local Addresses (this class of addresses has been formally deprecated but IANA has been requested not the reallocate the address range)
- FF00::/8—Multicast Addresses

11.3.1.1 Format of Aggregatable Global Unicast Addresses

The Aggregatable Global Unicast Address space is where the majority of hosts will initially find their home. Initially, it was further subdivided to create a hierarchy of allocation. Those subdivisions are identified in Figure 11.1.

- Public Topology
  - FP is the format prefix (always 001—i.e. 2000::/3).
  - TLA ID is the Top Level Aggregator Identifier.

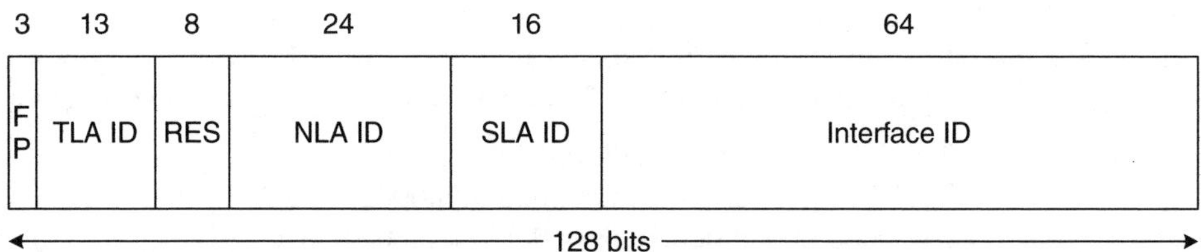

Figure 11.1 Original format of the globally aggregatable unicast addressing scheme

- RES is reserved for future use (always zero).
- NLA ID is the Next Level Aggregator Identifier.
- Site Topology
  - SLA ID is the Site Level Aggregator Identifier.
- Interface Identity
  - Interface ID is the Interface identifier.

However, this degree of formality in the network portion of the address was deemed too restrictive. Therefore, the aggregatable global unicast address space, currently 2000::/3, is simply broken up as in Figure 11.2.

- Public Topology
  - FP is the format prefix (currently 001—i.e. 2000::/3).
  - Global Routing Prefix.
- Site Topology
  - Subnet ID is allocatable by the holder of the /48 global prefix.
- Interface Identity
  - Interface ID is the Interface identifier.

When the first allocation of IPv6 addresses was made by IANA to the Regional Internet Registries (RIRs), it allocated only a single TLA, 0x0001, from within the FP 0x2000

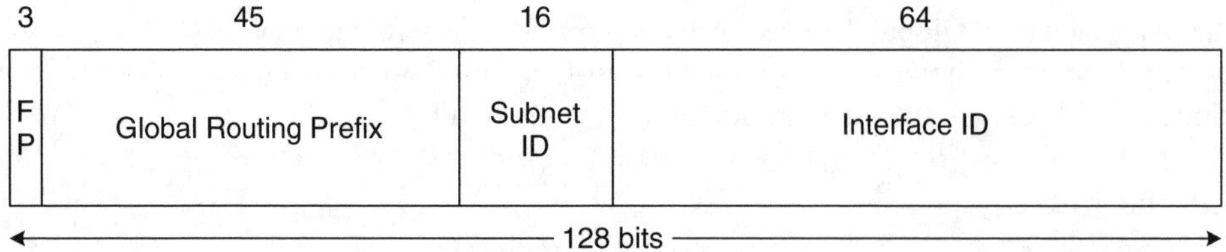

Figure 11.2 Newer format of the globally aggregatable unicast addressing scheme

giving the prefix 2001::0/16 in order to encourage a cautious allocation policy (and avoid a 'gold rush' for IPv6 addresses). This was divided between the RIRs, with each receiving an initial allocation of a /23. These could be further subdivided by the RIRs for allocation to the Local Internet Registries (LIR). The very first allocations to LIRs were /35s, however, the RIRs were free to allocate from their allocations in the manner they saw fit.

Every site is allocated a /48 prefix from which it could allocate up to 65536 subnets (16 bits of Subnet ID) in whatever hierarchical fashion it requires. Each of these subnets could, in theory, sustain up to 2^{64} hosts, although the problems associated with large flat subnets are just as valid with IPv6 as they are with IPv4. In terms of scaling, there is little justification for creating a single subnet with more than 64 bits allocated to the Interface ID. The main question about how to allocate the addresses relates to how multiple subnets might be aggregated at the numerous logical borders within the Internet. By carefully grouping the Subnet IDs, it is possible to significantly reduce the size of the routing table both globally and within the confines of your AS. This is a big issue with IPv4. With IPv6, it is going to be utterly essential to the successful growth of the Internet.

11.3.2 ADDRESS CLASSES

In addition to the functional categories listed above, as with IPv4, there are three classes of address in IPv6. However, unlike IPv4, which has unicast, multicast and broadcast, in IPv6 broadcast is considered simply a special case of multicast (i.e. a group to which all hosts belong).

- *Unicast Addresses.* As with IPv4, the main form of communication in IPv6 is between two hosts, i.e. unicast. The global aggregatable unicast address format has already been discussed in this section. With the exception of the addresses specifically allocated to multicast (see below), all addresses are considered to be unicast addresses.

- *Multicast Addresses.* Again, in common with IPv4, a block of addresses has been set aside specifically for use as multicast group addresses. IANA has allocated FF00::/8 for IPv6 multicast. Within that block, some sub-assignments have been made. FF01::/16 is used for node local scope, FF02::/16 for link local scope and FF05::/16 for site local scope. In addition, there are some 'well-known' permanently assigned addresses, which are valid across all scopes. Unsurprisingly, these are known as all-scope addresses. These are described in some detail in RFC 2375.

- *Anycast Addresses.* Given that one class of addresses has been discarded, it is clear that a new class of address must have been defined. Anycast addresses are used where a single global unicast address is assigned to more than one host. Each of the hosts (or its adjacent gateway/router) announces its address into the global routing table. It is then simply a case of selecting the topologically nearest of the available hosts bearing the same address. An example of the benefit of this mechanism is the geographical distribution of identical content. If a single address is advertised in the global routing table from several locations, each client will use the 'nearest' server to source the content. It is worth noting that anycast addresses are taken from the global unicast address space and are syntactically indistinguishable from any other unicast address.

Anycast addresses in IPv6 require a prefix length, which defines the topological area within which all the members of the anycast group are located. Within that area, the anycast address must be maintained as a host route. If, as might commonly be the case, the devices are distributed globally, then there is no way to constrain the scope of the host route. This is clearly going to limit the scalability and applicability of such a mechanism.

11.4   STATELESS AUTOCONFIGURATION

Manual configuration of addresses represented as 32 hexadecimal digits on every host in a network certainly presents an administrative headache. Such long addresses are prone to misconfiguration through simple typing errors. However, in this area IPv6 provides significant potential benefits over IPv4. The term stateless refers to the fact that no server needs to maintain any state with respect to the available networks or the range of valid host addresses. Instead, the interface address is derived from information already held by the interface (e.g. the MAC address) and the network prefix announced by on-net routers in their router advertisements. The routers sourcing the router advertisements clearly know the prefix because they have a manually configured interface address with an associated prefix length. This does not provide any obvious benefit for routers in the network. However, for hosts, it provides a significant benefit. The administrative load associated with managing large numbers of hosts attached to a network is significantly reduced. There is, of course, nothing to prevent the address of an interface being statically configured or the use of a stateful autoconfiguration mechanism like DHCP.

11.5   DOMAIN NAME SYSTEM (DNS)

As mentioned above, addresses represented as 32 hexadecimal digits are prone to mistyping and also, significantly, are more difficult to remember than the dotted quad format of IPv4 addresses. Therefore, DNS takes on a new importance with respect to providing human readable addresses for IPv6 interfaces. Names are mapped to IPv6 addresses using AAAA (pronounced 'quad A') records. The domain analogous to IPv4's in-addr.arpa for IPv6 is ip6.int. Reverse resolution of IPv6 addresses uses a very long format as demonstrated below:

1.0.0.0.0.0.0.0.0.0.0.0.0.0.0.0.0.0.0.0.F.E.D.C.B.A.9.8.7.6.5.4.3.2.1.ip6.int.

This is the reverse entry for 1234:5678:9ABC:DEF0:0000:0000:0000:0001.

11.6 TRANSITION MECHANISMS

It is clear that IPv6 is not going to supplant IPv4 in one hit on one 'flag day'. So, if IPv6 is to be deployed in more than just isolated pockets, there must be mechanisms to operate both networks in parallel and to provide links between IPv6 'islands' across the 'sea' of IPv4. There also must be a mechanism to provide connectivity between hosts using only IPv4 and those using only IPv6. Numerous mechanisms have been devised including dual-stack operation, Network Address Translation—Protocol Translation (NAT-PT) and IPv6 over various tunnelling mechanisms (e.g. IPv4, GRE, MPLS, etc.). Each has its benefits and disadvantages as described below.

11.6.1 DUAL STACK

This mode of operation requires each interface to be configured with both an IPv4 and an IPv6 address. On this basis, each incoming packet is handled by the appropriate IP stack for the type of packet. In this mode, both networks are run in parallel. This is similar to running both IPX and IP on a single network. Neither protocol interferes with, or even recognizes the presence of the other.

While this does nothing for the conservation of IP addresses, this is the preferred method of running both IPv4 and IPv6 in a contiguous network and is almost certainly the mode in which most SPs will be operated. It provides native support to both protocols, thus IPv4 hosts use IPv4 to talk to other IPv4 hosts and similarly IPv6 hosts use IPv6 to talk to other IPv6 hosts. One downside is that this provides no inherent mechanism for translating between the two protocols. So IPv4 hosts cannot communicate with IPv6 hosts unless some external gateway mechanism between the two networks is provided.

11.6.2 NETWORK ADDRESS TRANSLATION — PROTOCOL TRANSLATION

Having described the constraints of NAT, which have been used by many as a convincing argument for the development of IPv6, it seems somewhat ironic that one of the main mechanisms for joining the IPv4 and IPv6 worlds together is based on NAT-PT. NAT-PT is carried out by routers that support both IPv6 and IPv4. The fact that the hosts at each end of the conversation are not using the same version of IP is (or at least should be) totally opaque to the two hosts.

In addition to mapping the addresses in the headers of IP packets traversing the NAT-PT router, the router must also understand the protocols where IP addresses are carried within the data and, where appropriate, translate information within the data to match the mapping. For example, when a name server replies to a query from an IPv4-only host with an IPv6 address, the NAT-PT router must create the mapping between an IPv4 and the returned IPv6 address and then modify the DNS reply so that the IPv4-only host receives a DNS response containing IPv4 addresses.

Finally, the mapping of IPv4 header information to IPv6 header information is not a simple task. It is not always possible to translate completely from one to the other.

The requirement for such interrogation of every packet traversing the NAT-PT router makes this process exceptionally CPU intensive. Any such process will inevitably limit the scaling of the network.

For all these reasons, it is better to use other transition mechanisms where such mechanisms are available.

11.6.3　TUNNELLING IPv6 IN IPv4

There are several mechanisms based on tunnelling of one sort or another. They include:

- configured IPv6 in IPv4 tunnels;

- automatic IPv6 in IPv4 tunnels;

- IPv6 in MPLS.

11.6.3.1 Configured IPv6 in IPv4 Tunnels

This was the first mechanism widely deployed for the connection of disparate IPv6 networks and formed the basis of the experimental 6bone. This mechanism requires the static configuration of tunnels at both ends. This is administratively time-consuming and also somewhat inefficient (a problem associated with any tunnelling due to the extra overhead of the encapsulating protocol).

11.6.3.2 Automatic IPv6 in IPv4 Tunnels

In order to overcome some of the administrative burden of configuring IPv6 in IPv4 tunnels, a scheme has been developed to use IPv4 compatible addresses to automatically tunnel IPv6 addresses through an IPv4 network. IPv4 compatible IPv6 addresses embed an IPv4 address into an IPv6 address. The last 32 bits of the address is actually represented as a standard dotted quad address of the form:

$$1234:5678:9ABC:DEF0:0000:0000:192.168.103.112$$

11.6.3.3 IPv6 in MPLS

Just as MPLS is capable of transporting IPv4 packets across networks not capable of forwarding them without the MPLS encapsulation, it is similarly capable of transporting IPv6 packets. The encapsulation of IPv6 packets is identical to that for IPv4 packets with the exception that there is a different label for IPv6 explicit null (2).

11.7 **ROUTING IN IPv6**

In general, routing in IPv6 is no different from routing in IPv4. This is not entirely surprising since the forwarding of IPv6 packets is similar to the forwarding of IPv4 and the routing process is responsible for the selection of forwarding paths. Below, I have briefly described the differences between each of the routing protocols listed in Chapter 5 and their implementations for IPv6, if such an implementation exists. At the time of writing, IPv6 implementations of IS-IS, OSPF, RIPng and BGP have been created.

11.7.1 IS-IS FOR IPv6

As noted on several occasions before, IS-IS is, by design, extensible. It was extended to carry IPv4 NLRI. In order to carry IPv6 NLRI, IS-IS simply required the definition of some new TLVs. It is possible with a single IS-IS process to carry NLRI for IPv4, IPv6, CLNP, MPLS traffic engineering information and, theoretically, any other routed protocol simultaneously. This clearly provides the potential for enhanced scalability over any system in which routed protocols have to be carried in their own routing protocols. Because of the relative simplicity of extending IS-IS, it was the first widely deployed 'serious' IGP for IPv6.

11.7.2 OSPFv3

OSPFv3 is a new protocol designed explicitly for the exchange of IPv6 NLRI. It is not currently capable of exchanging IPv4 NLRI, although proposals are being put forward to enable OSPFv3 to carry multiple routed protocols in a single routing instance. This is potentially both a positive and a negative. On the negative side, having a separate routing process requires more memory and potentially more processing power to maintain the memory structures, routing state and associated overhead. On the positive side, if there is a failure in the routing protocol associated with one routed protocol (e.g. IPv4), having a different parallel routing protocol for another routed protocol (e.g. IPv6) would hopefully sustain that other routed protocol until such time as the failed routing protocol can be restored. This argument follows similar lines to those used against stateful failover of control modules (see Chapter 1).

11.7.3 RIPng

RIPng was the first routing protocol to be developed and widely deployed for IPv6. RIPng is, effectively, RIPv2 just extended to be able to carry IPv6 NLRI. It has no other enhancements or limitations beyond those associated with IPv4. Scaling issues are constrained to the extra size of the addresses being held in the routing table. Had it not been the first widely implemented routing protocol for IPv6, I suspect that RIPng would never have been particularly widely deployed.

11.7.4   MULTIPROTOCOL BGP

IPv6, as with IPv4, can use BGP to exchange unicast and multicast NLRI with neighbours, both within an autonomous system and beyond. Exchanging IPv6 NLRI is an extended capability that must be negotiated between peers. It is possible with most implementations to actually exchange IPv6 NLRI on a session between either IPv4 or IPv6 neighbours.

11.8   MULTICAST IN IPv6

- *Multicast Addressing.* As stated in Chapter 10, IPv6 uses particular blocks of addresses for particular functional elements. One such functional element is multicast and the address FF00::0/8 has been assigned to multicast.

- *Multicast Listener Discovery.* IPv6 does not use IGMP. Instead, a new protocol has been defined with similar functionality. This protocol is Multicast Listener Discovery (MLD). MLD, like IGMP, has already spawned two versions. Version 1 of MLD is somewhat analogous to IGMPv2, providing the host with the ability to identify the group it wishes to join. Version 2 of MLD is analogous to IGMPv3, which adds the ability to identify not only the group but also the source the host wishes to join.

- *Multicast Routing Protocols.* IPv6 uses the same multicast routing protocols as IPv4. The massive address space associated with IPv6 multicast addresses ($2^{128} - 1$ group addresses) exacerbates the problems experienced with multicast and IPv4. Dense mode routing protocols are likely to suffer from the problems associated with large numbers of groups repeatedly flooding and being pruned and the amount of state it is necessary to keep. For sparse mode protocols, specifically PIM-SM, scaling the RP presents a real problem with so many groups being sourced by a similarly vast number of sources. Source Specific Multicast (SSM) would certainly help in this case because it removes the need for the RP. However, using SSM also requires vast amounts of state to be stored on each device in the network. Since each source can transmit on $2^{128} - 1$ channels, the benefits of not having to maintain a RP are potentially overridden because every router has to deal with the same amount of state as the RP.

11.9   IPv6 SECURITY

IPsec was originally developed as an integral part of the IPv6 protocol. Authentication Headers (AH) and Encapsulating Security Payload (ESP) are standard headers in IPv6.

With IPsec as a built-in function of the underlying transport protocol, each host can effectively form security associations (SAs) with every other host with whom it wishes to communicate. Such one-to-one SAs would rely almost certainly on a PKI-based mechanism to authenticate each party to the other.

Since IPv6 provides sufficient addresses (for the foreseeable future) for every host to have a globally unique address, there is a greater chance that hosts will find themselves connected directly to the Internet, without the protection of combined firewalls/NAT boxes. While NAT, on its own, provides only a false sense of security, the combination of a well-designed and configured stateful firewall and NAT is a necessary security feature. Without that, hosts will inevitably be more vulnerable to attack or infection from viruses. Therefore, it is worth noting that while NAT will hopefully be a thing of the past, firewalls are here to stay.

In all other respects, the security issues affecting IPv6 are effectively identical to those affecting IPv4.

11.10 MOBILITY IN IPv6

In addition to the enhanced security features designed into IPv6, another feature designed into IPv6 and subsequently backported into IPv4 was IP Mobility. IPv6 Mobility allows nodes to move from one network to another, while maintaining lower layer connectivity. It relies upon a 'home gateway', which maintains knowledge of the actual location of the node and its home address. It is the home gateway that forwards traffic to and from the mobile node. Unlike IPv4 Mobility, IPv6 Mobility requires no 'foreign gateway' function.

On the subject of scaling, IPv6 Mobility has another benefit over IPv4 Mobility. Because it uses IPv6 extension headers and IPv6 encapsulation, it is not necessary for soft tunnel state to be maintained by the home gateway. If there are large numbers of regularly mobile hosts, the maintenance of soft state can be a significant drain on resources.

12

Complete Example Configuration Files (IOS and JUNOS Software)

In this chapter, we provide example configuration files for four major router types (a core router, a customer aggregation router, a peering router and a border router to an upstream transit provider). An example of a configuration file for each type of router is presented in both JUNOS and IOS. Every attempt has been made to ensure that the configurations using each configuration language are functionally identical. However, certain functionality is only supported by one or other of the vendors. In this case, the differences in functionality are highlighted in the annotations.

These configuration files should not be considered to be absolutely secure or absolutely functional for ever. For example, the filter protecting the routers is relatively simple and incomplete. However, it presents a template upon which you can build your own more comprehensive filters. Similarly, the list of martian prefixes is regularly updated as new prefixes are assigned to the Regional Internet Registries. See http://www.cymru.com/Bogons/index.html for a regularly updated list of martians (a.k.a bogons) and an automated mechanism for receiving those lists and integrating them into your routing policy. This site also provides comprehensive examples of secured JUNOS and IOS software configurations.

The configurations have comments to provide basic explanatory notes for each element. There are significant common elements to each configuration, which are only annotated in the first configuration in which they occur.

Designing and Developing Scalable IP Networks   G. Davies
© 2004 Guy Davies   ISBN: 0-470-86739-6

12.1 CORE ROUTER (P) RUNNING MPLS TE SUPPORTING LDP TUNNELLED THROUGH RSVP-TE, NO EDGE INTERFACES, IBGP ONLY, MULTICAST RP (ANYCAST STATIC) MSDP, PIM-SM (JUNOS)

```
/* Groups are used to create templates containing common information */
/* associated with multiple instances of a particular element of the
   configuration */
/* By applying the group, all matching elements are modified with the content
   */
/* of the group configuration */
groups {
    /* re0 and re1 are special groups used to create unique configuration */
    /* associated with the unit when each Routing Engine is the master */
    re0 {
        system {
            host-name London-Core1-re0;
        }
        interface {
            fxp0 {
                unit 0 {
                    family inet {
                        address 10.0.0.1/24;
                    }
                }
            }
        }
    }
    re1 {
        system {
            host-name London-Core1-re1;
        }
        interface {
            fxp0 {
                unit 0 {
                    family inet {
                        address 10.0.0.2/24;
                    }
                }
            }
        }
    }
    /* All SONET/SDH interfaces are configured to transport IS-IS and MPLS */
    sdh-template {
        interfaces {
            <so-*/*/*> {
                sonet-options {
                    rfc-2615;
                }
                family iso;
```

```
                    family mpls;
                }
            }
        }
        /* All LSPs have the same source and support LDP in RSVP tunneling */
        mpls-template {
            protocols {
                mpls {
                    label-switched-path <*> {
                        from 195.111.255.1;
                        ldp-tunneling;
                    }
                }
            }
        }
    }
}
apply-groups [ re0 re1 ];
system {
    domain-name exampledomain.com;
    time-zone Europe/London;
    /* disable protocol redirects on all interfaces */
    no-redirects;
    /* force core dumps to be created locally */
    dump-on-panic;
    /* force the contents of the Flash disk (primary boot media) */
    /* to be mirrored automatically onto the hard disk (secondary */
    /* boot media).  This should be disabled during upgrades since */
    /* it is mutually exclusive with "request system snapshot" */
    mirror-flash-on-disk;
    /* Configure a backup router to which this router sends packets */
    /* during startup and in the event of RPD failing */
    /* Ensure that this is constrained to the management network */
    backup-router 10.0.0.254 destination 10.0.0.0/8;
    /* use RADIUS for authentication */
    /* use local entries only if RADIUS server not available */
    authentication-order radius;
    /* set the console port parameters */
    ports {
        console type vt100;
    }
    /* automatically save the config daily and on each commit */
    archival configuration {
        transfer-interval 1440;
        transfer-on-commit;
        archive-sites {
            ftp://robot-user:robot-pwd@195.111.253.8/data/configs;
        }
    }
    /* configure a strong root password */
    root-authentication {
        encrypted-password "<encrypted password>"; # SECRET-DATA
    }
    /* configure a minimum of two nameservers */
    name-server {
```

```
            195.111.253.1;
            195.111.254.1;
        }
        /* configure a minimum of two radius servers */
        radius-server {
            195.111.253.2 {
                secret "<encrypted secret key>"; # SECRET-DATA
                timeout 1;
                retry 2;
            }
            195.111.254.2 {
                secret "<encrypted secret key>"; # SECRET-DATA
                timeout 1;
                retry 2;
            }
        }
        /* user and class configuration */
        login {
            /* the login message has legal status in some jurisdictions */
            message "\n         +-------------------------------------+ \n    |
|\n        | ********* W A R N I N G ********* |\n         |
|\n        |   STRICTLY NO UNAUTHORISED ACCESS   |\n         |
|\n        +-------------------------------------+\n\n";
            /* a local superuser class - the default cannot be modified */
            class mysuperuser {
                idle-timeout 60;
                permissions all;
            }
            /* a class for network operators */
            class operator {
                permissions [ admin configure interface interface-control network
                    routing snmp system trace view firewall ];
                allow-commands "clear interfaces statistics";
            }
            /* a read-only class */
            class viewall {
                idle-timeout 60;
                permissions [ admin interface network routing snmp system trace
                    view firewall ];
            }
            user mysuperuser {
                full-name "Template RADIUS Account";
                uid 1000;
                class mysuperuser;
            }
            user operator {
                full-name "Template RADIUS Account";
                uid 1001;
                class operator;
            }
            user viewall {
                full-name "Template RADIUS Account";
                uid 1002;
```

```
            class viewall;
        }
        /* create a last resort local login account */
        user admin {
            full-name "Last resort account";
            uid 9999;
            class mysuperuser;
            authentication {
                encrypted-password "<encrypted password>";
            }
        }
    }
    services {
        /* permit ftp only when specifically needed.  Otherwise leave
           deactivated */
        inactive: ftp {
            connection-limit 1;
            rate-limit 1;
        }
        /* permit only SSHv2 authentication */
        ssh {
            root-login deny;
            protocol-version 2;
            connection-limit 4;
            rate-limit 4;
        }
        /* permit telnet only if SSH is not available - e.g. export releases
           */
        inactive: telnet {
            connection-limit 4;
            rate-limit 4;
        }
    }
    /* configure system logging */
    syslog {
        archive size 10m files 10;
        user * {
            any emergency;
        }
        host 195.111.253.3 {
            any info;
        }
        host 195.111.254.3 {
            any info;
        }
        file messages {
            any info;
            authorization info;
        }
        file security.log {
            interactive-commands any;
            archive size 10m files 10;
        }
    }
```

```
        /* configure at least two ntp servers */
        ntp {
            /* boot-server will force the time to be set from */
            /* the server each time the router is reloaded */
            boot-server 195.111.253.4;
            server 195.111.253.4;
            server 195.111.254.4;
        }
    }
    chassis {
        no-reset-on-timeout;
        /* Dump core files to the local hard drive */
        dump-on-panic;
        /* Configure the redundancy features including the failover between two
            REs */
        redundancy {
            routing-engine 0 master;
            routing-engine 1 backup;
            failover on-loss-of-keepalives;
            keepalive-time 300;
        }
        /* Configure all POS interfaces to use SDH rather than the default SONET
            */
        fpc 0 {
            pic 0 {
                framing sdh;
            }
        }
        fpc 1 {
            pic 0 {
                framing sdh;
            }
        }
        fpc 6 {
            pic 0 {
                framing sdh;
            }
        }
        fpc 7 {
            pic 0 {
                framing sdh;
            }
        }
    }
    interfaces {
        apply-groups sdh-template;
        so-0/0/0 {
            description "Link to London-Core2";
            clocking internal;
            encapsulation ppp;
            unit 0 {
                description "Link to London-Core2 POS0/0";
                family inet {
                    address 195.111.1.1/30;
```

```
                }
            }
        }
        so-1/0/0 {
            description "Link to Frankfurt-Core1";
            clocking external;
            encapsulation ppp;
            unit 0 {
                description "Link to Frankfurt-Core1 so-1/0/0.0";
                family inet {
                    address 195.111.1.5/30;
                }
            }
        }
        ge-2/0/0 {
            description "GigE to primary local LAN";
            unit 0 {
                family inet {
                    address 195.111.100.1/24;
                }
                family iso;
                family mpls;
            }
        }
        ge-3/0/0 {
            description "GigE to backup local LAN";
            unit 0 {
                family inet {
                    address 195.111.101.1/24;
                }
                family iso;
                family mpls;
            }
        }
        so-6/0/0 {
            description "Link to SanFrancisco-Core1";
            clocking internal;
            encapsulation ppp;
            unit 0 {
                description "Link to SanFrancisco-Core1 so-6/0/0.0";
                family inet {
                    address 195.111.1.9/30;
                }
            }
        }
        so-7/0/0 {
            description "Link to Boston-Core1";
            enable;
            clocking internal;
            unit 0 {
                description "Link to Boston-Core1 so-7/0/0.0";
                family inet {
                    address 195.111.1.13/30;
                }
```

```
        }
    }
    lo0 {
        unit 0 {
            family inet {
                /* filtering on the loopback protects the routing engine from
                   */
                /* traffic arriving on all interfaces */
                filter {
                    input accessfilter;
                }
                address 195.111.255.1/32 {
                    primary;
                    preferred;
                }
                address 195.111.255.254/32;
            }
            family iso {
                address 49.0000.1921.6825.5001.00;
            }
        }
    }
}
forwarding-options {
    /* This configures the CflowD format sampling */
    sampling {
        input {
            family inet {
                rate 500;
            }
        }
        output {
            cflowd 195.111.254.5 {
                version 5;
                local-dump;
                port 9995;
                autonomous-system-type origin;
            }
        }
    }
}
snmp {
    description "London-Core1 Juniper M160";
    location "London";
    contact "Netops +44 1999 999999";
    community <string> {
        authorization read-only;
        clients {
            /* List each client individually */
            195.111.253.5/32;
            195.111.253.6/32;
        }
    }
    trap-options {
```

```
            source-address lo0;
        }
        trap-group Juniper {
            version v2;
            categories {
                authentication;
                chassis;
                link;
                routing;
                startup;
            }
            targets {
                195.111.253.5;
                195.111.253.6;
            }
        }
    }
routing-options {
    static {
        /* route to Out of Band Network */
        route 10.0.0.0/8 {
            next-hop 10.0.0.254;
            retain;
            no-readvertise;
        }
    }
    aggregate {
        route 195.111.0.0/16;
    }
    autonomous-system 1;
    forwarding-table {
        /* enable per-flow load sharing */
        /* see the policy-statement load-share for more details */
        export load-share;
    }
    interface-routes {
        rib-group inet interface-rib;
    }
    rib-groups {
        bgp-rib {
            import-rib [ inet.0 inet.2 ];
            export-rib [ inet.0 ];
        }
        multicast-rib {
            import-rib [ inet.2 ];
            export-rib [ inet.2 ];
        }
        interface-rib {
            import-rib [ inet.0 inet.2 ];
        }
    }
}
protocols {
    /* This is actually RSVP-TE */
```

```
rsvp {
    traceoptions {
        file rsvp.log files 10 size 1m;
        flag hello detail;
        flag state detail;
    }
    log-updown;
    /* enable RSVP on all core facing interfaces */
    interface so-0/0/0.0;
    interface so-1/0/0.0;
    interface so-6/0/0.0;
    interface so-7/0/0.0;
    /* enable RSVP on the loopback interface */
    interface lo0.0;
    /* explicitly disable RSVP on the management interface */
    interface fxp0.0 {
        disable;
    }
}
mpls {
    traceoptions {
        file mpls.log files 10 size 1m;
        flag hello detail;
        flag state detail;
    }
    /* Apply the template defined under the groups section */
    apply-groups mpls-template;
    log-updown;
    /* Mask the internal architecture of the network */
    no-propagate-ttl;
    traffic-engineering bgp;
    /* Configure the destination end point for all LSPs */
    /* The source end point and other parameters are inherited from the
        */
    /* mpls-template */
    label-switched-path LonP1-LonP2 {
        to 195.111.255.2;
    }
    label-switched-path LonP1-FraP1 {
        to 195.111.255.3;
    }
    label-switched-path LonP1-FraP2 {
        to 195.111.255.4;
    }
    label-switched-path LonP1-BosP1 {
        to 195.111.255.5;
    }
    label-switched-path LonP1-BosP2 {
        to 195.111.255.6;
    }
    label-switched-path LonP1-SFrP1 {
        to 195.111.255.7;
    }
    label-switched-path LonP1-SFrP2 {
```

```
                to 195.111.255.8;
        }
        /* Enable MPLS on all configured interfaces */
        interface so-0/0/0.0;
        interface so-1/0/0.0;
        interface ge-2/0/0.0;
        interface ge-3/0/0.0;
        interface so-6/0/0.0;
        interface so-7/0/0.0;
        interface lo0.0;
        /* Explicitly disable MPLS on the management interface */
        interface fxp0.0 {
            disable;
        }
    }
    ldp {
        traceoptions {
            file ldp.log files 10 size 1m;
            flag hello detail;
            flag state detail;
        }
        log-updown;
        /* Enable LDP on the edge facing interfaces */
        interface ge-2/0/0.0;
        interface ge-3/0/0.0;
        /* Enable LDP on the loopback interface */
        interface lo0.0;
        /* Explicitly disable LDP on the management interface */
        interface fxp0.0 {
            disable;
        }
    }
    bgp {
        traceoptions {
            file bgp.log files 10 size 1m;
            flag open detail;
            flag state detail;
        }
        /* Enable BGP to inject routing information into both unicast and */
        /* multicast routing tables */
        rib-group bgp-rib;
        log-updown;
        export our-aggregates;
        /* A peer group for Core routers */
        group CORE {
            type internal;
            local-address 195.111.255.1;
            neighbor 195.111.255.2; /* LonP2 */
            neighbor 195.111.255.3; /* FraP1 */
            neighbor 195.111.255.4; /* FraP2 */
            neighbor 195.111.255.5; /* BosP1 */
            neighbor 195.111.255.6; /* BosP2 */
            neighbor 195.111.255.7; /* SFrP1 */
            neighbor 195.111.255.8; /* SFrP2 */
```

```
        }
        /* A Route-Reflection peer group for locally connected edge routers
           */
        group LONDON-RR-CLUSTER {
            type internal;
            local-address 195.111.255.1;
            cluster 195.111.255.1;
            neighbor 195.111.255.100; /* Access1 */
            neighbor 195.111.255.200; /* Peer1 */
            neighbor 195.111.255.210; /* Transit1 */
        }
    }
    isis {
        traceoptions {
            file isis.log files 10 size 1m;
            flag hello detail;
            flag error detail;
            flag state detail;
        }
        /* Enable IS-IS to populate both unicast and multicast routing tables
           */
        multicast-topology;
        /* Enable IS-IS on each configured interface and specify the metrics
           */
        /* at Level 1 and Level 2 */
        /* Disable Level 2 on non-core links */
        interface so-0/0/0.0 {
            level 1 metric 20;
            level 2 metric 40;
        }
        interface so-1/0/0.0 {
            level 1 disable;
            level 2 metric 55;
        }
        interface ge-2/0/0.0 {
            level 1 {
                metric 25;
                hello-interval 10;
                hold-time 40;
            }
            level 2 disable;
        }
        interface ge-3/0/0.0 {
            level 1 {
                metric 30;
                hello-interval 10;
                hold-time 40;
            }
            level 2 disable;
        }
        interface so-6/0/0.0 {
            level 1 disable;
            level 2 metric 10;
        }
```

```
        interface so-7/0/0.0 {
            level 1 disable;
            level 2 metric 10;
        }
        interface lo0.0;
        interface fxp0.0 {
            disable;
        }
    }
    pim {
        traceoptions {
            file pim.log files 10 size 1m;
            flag hello detail;
            flag join detail;
            flag rp detail;
            flag assert detail;
            flag state detail;
        }
        /* Enable PIM to install and use information in the multicast routing
           table */
        rib-group multicast-rib;
        /* Configure the Rendezvous-Point using a statically configured */
        /* anycast address */
        rp {
            local {
                family inet {
                    address 195.111.255.254;
                }
            }
        }
        /* Enable PIM sparse mode on all configured interfaces */
        interface so-0/0/0.0 {
            mode sparse;
        }
        interface so-1/0/0.0{
            mode sparse;
        }
        interface ge-2/0/0.0{
            mode sparse;
        }
        interface ge-3/0/0.0{
            mode sparse;
        }
        interface so-6/0/0.0{
            mode sparse;
        }
        interface so-7/0/0.0{
            mode sparse;
        }
        interface lo0.0{
            mode sparse;
        }
        /* Explicitly disable PIM on the management interface */
        interface fxp0.0 {
```

```
                disable;
            }
        }
        msdp {
            traceoptions {
                file msdp.log files 10 size 1m;
                flag hello detail;
                flag state detail;
            }
            log-updown;
            /* Enable MSDP to install and use information in the m'cast routing
               table */
            rib-group multicast-rib;
            local-address 195.111.255.1;
            /* Configure each of the other RPs to be neighbours */
            group rp-group {
                peer 195.111.255.3; /* FraP1 */
                peer 195.111.255.5; /* BosP1 */
                peer 195.111.255.7; /* SFrP1 */
            }
        }
    }
    policy-options {
        policy-statement load-share {
            term load-balance-per-flow {
                then {
                    /* this is slightly misleading.  In an IP II based
                       architecture */
                    /* this actually causes per-flow load sharing */
                    load-balance per-packet;
                }
            }
        }
        policy-statement our-aggregates {
            term our-aggregates {
                from {
                    protocol aggregate;
                    route-filter 195.111.0.0/16 exact;
                }
                then {
                    community set OurAggregates;
                    accept;
                }
            }
        }
        community OurAggregates members 1:50000;
    }
    class-of-service {
        classifiers {
            /* Configure a classifier based on MPLS EXP bits */
            exp mpls-exp {
                forwarding-class expedited-forwarding {
                    loss-priority high code-points [ 101 110 111 ];
                }
```

```
                forwarding-class assured-forwarding {
                    loss-priority high code-points 011;
                }
                forwarding-class best-effort {
                    loss-priority high code-points [ 000 010 100 ];
                }
                forwarding-class less-than-best-effort {
                    loss-priority high code-points 001;
                }
            }
            /* Configure a classifier based on DiffServ Code Points */
            dscp diffserv {
                forwarding-class expedited-forwarding {
                    loss-priority high code-points [ ef cs6 cs7 cs5 ];
                }
                forwarding-class assured-forwarding {
                    loss-priority high code-points af31;
                    loss-priority low code-points [ af32 af33 ];
                }
                forwarding-class best-effort {
                    loss-priority high code-points [ be cs2 cs4 ];
                }
                forwarding-class less-than-best-effort {
                    loss-priority high code-points cs1;
                }
            }
            /* Configure a classifier based on IP precedence in the ToS byte */
            inet-precedence inet-prec {
                forwarding-class expedited-forwarding {
                    loss-priority high code-points [ 101 110 111 ];
                }
                forwarding-class assured-forwarding {
                    loss-priority high code-points 011;
                }
                forwarding-class best-effort {
                    loss-priority high code-points [ 000 010 100 ];
                }
                forwarding-class less-than-best-effort {
                    loss-priority high code-points 001;
                }
            }
        }
    /* Configure two RED drop profiles, one aggressive and one gentle */
    drop-profiles {
        aggressive {
            fill-level 10 drop-probability 30;
            fill-level 70 drop-probability 100;
        }
        gentle {
            fill-level 20 drop-probability 20;
            fill-level 80 drop-probability 80;
        }
    }
    /* Assign forwarding-class names to the four available queues */
```

```
forwarding-classes {
    queue 0 best-effort;
    queue 1 less-than-best-effort;
    queue 2 assured-forwarding;
    queue 3 expedited-forwarding;
}
/* Associate classifiers (for inbound traffic) and schedulers */
/* (for outbound traffic) to each interface */
interfaces {
    so-0/*/* {
        scheduler-map diffserv;
        unit * {
            classifiers {
                dscp diffserv;
                exp mpls-exp;
            }
        }
    }
    so-1/*/* {
        scheduler-map diffserv;
        unit * {
            classifiers {
                dscp diffserv;
                exp mpls-exp;
            }
        }
    }
    ge-2/*/* {
        scheduler-map diffserv;
        unit * {
            classifiers {
                dscp diffserv;
                exp mpls-exp;
            }
        }
    }
    ge-3/*/* {
        scheduler-map diffserv;
        unit * {
            classifiers {
                dscp diffserv;
                exp mpls-exp;
            }
        }
    }
    so-6/*/* {
        scheduler-map diffserv;
        unit * {
            classifiers {
                dscp diffserv;
                exp mpls-exp;
            }
        }
    }
```

```
        so-7/*/* {
            scheduler-map diffserv;
            unit * {
                classifiers {
                    dscp diffserv;
                    exp mpls-exp;
                }
            }
        }
    }
    /* Associate a scheduler with a forwarding-class */
    scheduler-maps {
        diffserv {
            forwarding-class expedited-forwarding scheduler
                expedited-forwarding;
            forwarding-class assured-forwarding scheduler assured-forwarding;
            forwarding-class best-effort scheduler best-effort;
            forwarding-class less-than-best-effort scheduler
                less-than-best-effort;
        }
    }
    /* Define schedulers based on the percentage of total service, buffers */
    /* and the RED drop profile */
    schedulers {
        expedited-forwarding {
            transmit-rate percent 10 exact;
            buffer-size percent 10;
            priority strict-high;
        }
        assured-forwarding {
            transmit-rate percent 10;
            buffer-size percent 10;
            priority high;
            drop-profile-map loss-priority high protocol any drop-profile
                aggressive;
            drop-profile-map loss-priority low protocol any drop-profile
                gentle;
        }
        best-effort {
            transmit-rate percent 79;
            buffer-size percent 79;
            priority low;
            drop-profile-map loss-priority any protocol any drop-profile
                gentle;
        }
        less-than-best-effort {
            transmit-rate remainder;
            buffer-size percent 1;
            priority low;
            drop-profile-map loss-priority any protocol any drop-profile
                gentle;
        }
    }
}
```

```
firewall {
    /* Define a filter to be applied to traffic destined for this device */
    filter accessfilter {
        term allow-telnet-from-very-small-list {
            from {
                source-address {
                    /* local hosts */
                    195.111.250.0/24;
                    /* local routers */
                    195.111.255.0/24;
                }
                protocol tcp;
                destination-port telnet;
            }
            then accept;
        }
        term deny-telnet-from-everywhere-else {
            from {
                protocol tcp;
                destination-port telnet;
            }
            then {
                log;
                reject;
            }
        }
        term allow-ftp-from-very-small-list {
            from {
                source-address {
                    /* local hosts */
                    195.111.250.0/24;
                }
                protocol tcp;
                port [ ftp ftp-data ];
            }
            then accept;
        }
        term deny-ftp-from-everywhere-else {
            from {
                protocol tcp;
                destination-port [ ftp ftp-data ];
            }
            then {
                log;
                reject;
            }
        }
        term allow-ssh-from-small-list {
            from {
                source-address {
                    /* local hosts */
                    195.111.250.0/24;
                    /* local routers */
                    195.111.255.0/24;
```

```
                    /* partner network */
                    10.24.0.0/24;
                }
                protocol tcp;
                destination-port ssh;
            }
            then accept;
        }
        term deny-ssh-from-everywhere-else {
            from {
                protocol tcp;
                destination-port ssh;
            }
            then {
                log;
                reject;
            }
        }
        term allow-everything-else {
            then accept;
        }
    }
    /* Configure Cflow sampling */
    filter cflowd {
        term sample-and-accept {
            then {
                sample;
                accept;
            }
        }
    }
}
```

12.2   CORE ROUTER (P) RUNNING MPLS TE SUPPORTING LDP TUNNELLED THROUGH RSVP-TE, NO EDGE INTERFACES, IBGP ONLY, MULTICAST RP (ANYCAST STATIC) MSDP, PIM-SM (IOS)

```
no service tcp-small-servers
no service udp-small-servers
service password-encryption
no service pad
no ip finger
no ip bootp server
!
! Configure AAA to use RADIUS and default to local authentication only if
```

```
! the RADIUS server does not respond.  Also configure accounting of commands
    and
! connections
aaa new-model
aaa authentication login use-radius group radius local
aaa authorization exec default group radius
aaa accounting commands 0 default start-stop group radius
aaa accounting system default start-stop group radius
aaa accounting connection default start-stop group radius
!
enable secret 5 XXXXXXXXX
!
! Configure an administrator password that will only be used if the RADIUS
    server does
! not respond
username admin password 7 XXXXXXXXX
hostname London-Core1
ip domain-name exampledomain.com
ip name-server 195.111.253.1 195.111.254.1
!
! Configure the router to dump corefiles to an ftp server using the user
    "robot-user"
ip ftp username robot-user
ip ftp password 0 robot-pwd
exception protocol ftp
exception dump 195.111.253.8
exception core-file London-Core1-corefile
!
! Enable dCEF (if it is not enabled by default)
! CEF is required for MPLS
ip cef distributed
!
! Enable MPLS
mpls ip
! Constrain the labels being advertised to those in the loopback range
mpls ldp advertise-labels for 1
! Define the router-id to be the address of the loopback0 interface
mpls ldp router-id lo0
! Enable targeted hellos (required to transport LDP through RSVP)
mpls ldp discovery targeted-hellos accept
!
! Enable MPLS traffic engineering using RSVP-TE
mpls traffic-eng tunnels
!
! Enable multicast routing
ip multicast-routing
!
! Configure Syslog servers
logging 195.111.253.3 6
logging 195.111.254.3 6
logging console 0
logging source-interface lo0
!
! Configure NetFlow
```

```
ip flow-export 195.111.254.5 9995 version 5 origin-as
!
! Configure Interfaces
interface POS0/0
 description POS link to London-Core2 POS0/0
 ip address 195.111.1.1 255.255.255.252
 ! Apply an access filter to all interfaces to protect this device
 ip access-group accessfilter in
 ! Configure the POS interface to use SDH (rather than the default SONET)
 pos framing sdh
 pos flag s1s0 2
 pos scramble-atm
 crc 32
 ! Configure the device to use the internal clock to synchronize transmission
   on
 ! the POS link
 clock source internal
 ! Apply QoS to this interface
 service-policy output core-forwarding
 ! Enable and configure IS-IS on this interface
 ip router isis
 isis circuit-type level-1-2
 isis metric 20 level-1
 isis metric 40 level-2
 ! Enable and configure MPLS on this interface
 mpls ip
 mpls traffic-eng tunnels
 ! Enable multicast on this interface
 ip pim sparse-mode
!
interface POS1/0
 description POS link to Frankfurt-Core1 POS1/0
 ip address 195.111.1.5 255.255.255.252
 ip access-group accessfilter in
 pos framing sdh
 pos flag s1s0 2
 pos scramble-atm
 crc 32
 ! Configure this interface to use a remote clock source (either from the
   transmission
 ! equipment or the remote router
 clock source line
 service-policy output core-forwarding
 ip router isis
 isis circuit-type level-2-only
 isis metric 55 level-2
 mpls ip
 mpls traffic-eng tunnels
 ip pim sparse-mode
!
interface GigabitEthernet2/0
 description GigE to primary local LAN
 ip address 195.111.100.1 255.255.255.0
 ip access-group accessfilter in
```

```
 service-policy output core-forwarding
 ip router isis
 isis circuit-type level-1
 isis metric 25 level-1
 isis hello-interval 10 level-1
 isis hello-multiplier 4
 mpls ip
 mpls label protocol ldp
 ip pim sparse-mode
!
interface GigabitEthernet3/0
 description GigE to secondary local LAN
 ip address 195.111.101.1 255.255.255.0
 ip access-group accessfilter in
 service-policy output core-forwarding
 ip router isis
 isis circuit-type level-1
 isis metric 30 level-1
 isis hello-interval 10 level-1
 isis hello-multiplier 4
 mpls ip
 mpls label protocol ldp
 ip pim sparse-mode
!
interface Ethernet 4/0/0
 description Link to Management Network
 ip address 10.0.0.1 255.255.255.0
 ip access-group mgmt-in in
 ip access-group mgmt-out out
!
interface POS6/0
 description POS link to SanFrancisco-Core1 POS6/0
 ip address 195.111.1.9 255.255.255.252
 ip access-group accessfilter in
 pos framing sdh
 pos flag s1s0 2
 pos scramble-atm
 crc 32
 clock source line
 service-policy output core-forwarding
 ip router isis
 isis circuit-type level-2-only
 isis metric 55 level-2
 mpls ip
 mpls traffic-eng tunnels
 ip pim sparse-mode
!
interface POS7/0
 description POS link to Boston-Core1 POS7/0
 ip address 195.111.1.13 255.255.255.252
 ip access-group accessfilter in
 pos framing sdh
 pos flag s1s0 2
 pos scramble-atm
```

```
 crc 32
 clock source line
 service-policy output core-forwarding
 ip router isis
 isis circuit-type level-2-only
 isis metric 55 level-2
 mpls ip
 mpls traffic-eng tunnels
 ip pim sparse-mode
!
interface lo0
 description Loopback
 ip address 195.111.255.1 255.255.255.255
 ip router isis
!
interface lo1
 description PIM Anycast RP address
 ip address 195.111.255.254 255.255.255.255
 ip router isis
 ip pim sparse-mode
!
! Configure MPLS TE tunnels
interface tunnel2
 description MPLS TE tunnel LonP1-LonP2
 ip unnumbered lo0
 tunnel source 195.111.255.1
 tunnel destination 195.111.255.2
 tunnel mode mpls traffic-eng
 tunnel mpls traffic-eng path-option 1 dynamic
 tunnel mpls traffic-eng autoroute announce
 mpls ip
 mpls label protocol ldp
!
interface tunnel3
 description MPLS TE tunnel LonP1-FraP1
 ip unnumbered lo0
 tunnel source 195.111.255.1
 tunnel destination 195.111.255.3
 tunnel mode mpls traffic-eng
 tunnel mpls traffic-eng path-option 1 dynamic
 tunnel mpls traffic-eng autoroute announce
 mpls ip
 mpls label protocol ldp
!
interface tunnel4
 description MPLS TE tunnel LonP1-FraP2
 ip unnumbered lo0
 tunnel source 195.111.255.1
 tunnel destination 195.111.255.4
 tunnel mode mpls traffic-eng
 tunnel mpls traffic-eng path-option 1 dynamic
 tunnel mpls traffic-eng autoroute announce
 mpls ip
 mpls label protocol ldp
```

```
!
interface tunnel5
 description MPLS TE tunnel LonP1-BosP1
 ip unnumbered lo0
 tunnel source 195.111.255.1
 tunnel destination 195.111.255.5
 tunnel mode mpls traffic-eng
 tunnel mpls traffic-eng path-option 1 dynamic
 tunnel mpls traffic-eng autoroute announce
 mpls ip
 mpls label protocol ldp
!
interface tunnel6
 description MPLS TE tunnel LonP1-BosP2
 ip unnumbered lo0
 tunnel source 195.111.255.1
 tunnel destination 195.111.255.6
 tunnel mode mpls traffic-eng
 tunnel mpls traffic-eng path-option 1 dynamic
 tunnel mpls traffic-eng autoroute announce
 mpls ip
 mpls label protocol ldp
!
interface tunnel7
 description MPLS TE tunnel LonP1-SfrP1
 ip unnumbered lo0
 tunnel source 195.111.255.1
 tunnel destination 195.111.255.7
 tunnel mode mpls traffic-eng
 tunnel mpls traffic-eng path-option 1 dynamic
 tunnel mpls traffic-eng autoroute announce
 mpls ip
 mpls label protocol ldp
!
interface tunnel8
 description MPLS TE tunnel LonP1-SfrP2
 ip unnumbered lo0
 tunnel source 195.111.255.1
 tunnel destination 195.111.255.8
 tunnel mode mpls traffic-eng
 tunnel mpls traffic-eng path-option 1 dynamic
 tunnel mpls traffic-eng autoroute announce
 mpls ip
 mpls label protocol ldp
!
! Create a static route to null0 for aggregate route
ip route 195.111.0.0 255.255.0.0 null0
!
! All loopback 0 addresses
ip access-list 1 permit 195.111.255.0 0.0.0.255
! All SNMP clients
ip access-list 99 permit host 195.111.253.5
ip access-list 99 permit host 195.111.253.6
!
```

```
! All management networks permitted to talk to the loopback and management
   interface
ip access-list extended mgmt-in
 permit 10.0.0.0 0.0.0.255 host 195.111.255.1
 permit 10.0.0.0 0.0.0.255 host 10.0.0.1
 deny any any
! Local Loopback and Management interface permitted to talk to management
   networks
ip access-list extended mgmt-out
 permit host 195.111.255.1 10.0.0.0 0.255.255.255
 permit host 10.0.0.1 10.0.0.0 0.255.255.255
 deny any any
! An access-list to protect this device
ip access-list extended accessfilter
 permit ip 195.111.250.0 0.0.0.255 host 195.111.255.1 eq ssh
 permit ip 195.111.250.0 0.0.0.255 195.111.0.0 0.0.3.255 eq ssh
 permit ip 195.111.255.0 0.0.0.255 host 195.111.255.1 eq ssh
 permit ip 195.111.255.0 0.0.0.255 195.111.0.0 0.0.3.255 eq ssh
 permit ip 10.24.0.0 0.0.0.255 host 195.111.255.1 eq ssh
 permit ip 10.24.0.0 0.0.0.255 195.111.0.0 0.0.3.255 eq ssh
 deny ip any host 195.111.255.1 eq ssh
 deny ip any 195.111.0.0 0.0.3.255 eq ssh
 permit ip 195.111.250.0 0.0.0.255 host 195.111.255.1 eq ftp
 permit ip 195.111.250.0 0.0.0.255 host 195.111.255.1 eq ftp-data
 permit ip 195.111.255.0 0.0.0.255 host 195.111.255.1 eq ftp
 permit ip 195.111.255.0 0.0.0.255 host 195.111.255.1 eq ftp-data
 permit ip 195.111.250.0 0.0.0.255 195.111.0.0 0.0.3.255 eq ftp
 permit ip 195.111.250.0 0.0.0.255 195.111.0.0 0.0.3.255 eq ftp-data
 permit ip 195.111.255.0 0.0.0.255 195.111.0.0 0.0.3.255 eq ftp
 permit ip 195.111.255.0 0.0.0.255 195.111.0.0 0.0.3.255 eq ftp-data
 deny ip any host 195.111.255.1 eq ftp
 deny ip any host 195.111.255.1 eq ftp-data
 deny ip any 195.111.0.0 0.0.3.255 eq ftp
 deny ip any 195.111.0.0 0.0.3.255 eq ftp
 permit ip 195.111.250.0 0.0.0.255 host 195.111.255.1 eq telnet
 permit ip 195.111.250.0 0.0.0.255 195.111.0.0 0.0.3.255 eq telnet
 permit ip 195.111.255.0 0.0.0.255 host 195.111.255.1 eq telnet
 permit ip 195.111.255.0 0.0.0.255 195.111.0.0 0.0.3.255 eq telnet
 deny ip any host 195.111.255.1 eq ssh
 deny ip any 195.111.0.0 0.0.3.255 eq ssh
 permit ip any any
! Our Aggregates
ip access-list standard our-aggregates
 permit 195.111.0.0 0.0.255.255
!
ip community-list standard OurAggregates permit 1:10000
!
! QoS classifiers for IP precedence, MPLS EXP and DiffServ Code Points
class-map match-any expedited-forwarding
 match ip precedence 101 110 111
 match mpls experimental 101 110 111
 match ip dscp ef cs5 cs6 cs7
class-map match-any assured-forwarding
 match ip precedence 011
```

```
 match mpls experimental 011
 match ip dscp af31 af32 af33
class-map match-any best-effort
 match ip precedence 000 010 100
 match mpls experimental 000 010 100
 match ip dscp default cs2 cs4
class-map match-any less-than-best-effort
 match ip precedence 001
 match mpls experimental 001
 match ip dscp cs1
!
! QoS forwarding schedulers
policy-map core-forwarding
 class expedited-forwarding
 bandwidth percent 10
 class assured-forwarding
 bandwidth percent 10
 class best-effort
 bandwidth percent 79
 class less-than-best-effort
 bandwidth percent 1
!
! IS-IS configuration
router isis
 net 49.0000.1921.6825.5001.00
 is-type level-1-2
 ! Enable IS-IS TE extensions
 mpls traffic-eng level-2
 mpls traffic-eng router-id 195.111.255.1
 metric-style wide
!
! BGP configuration
router bgp 1
 no synchronize
 ! Redistribute the aggregate route into BGP
 redistribute static route-map our-aggregates
 ! Peer group for core routers
 neighbor CORE peer-group
 neighbor CORE remote-as 1
 neighbor CORE send-community
 neighbor CORE update-source loopback0
 ! Route-reflection cluster for local edge routers
 neighbor LONDON-RR-CLUSTER peer-group
 neighbor LONDON-RR-CLUSTER remote-as 1
 neighbor LONDON-RR-CLUSTER send-community
 neighbor LONDON-RR-CLUSTER update-source loopback0
 neighbor LONDON-RR-CLUSTER cluster-id 195.111.255.1
 neighbor LONDON-RR-CLUSTER reflector-client
 ! Associate specific neighbours with peer-groups
 neighbor 195.111.255.2 peer-group CORE
 neighbor 195.111.255.3 peer-group CORE
 neighbor 195.111.255.4 peer-group CORE
 neighbor 195.111.255.5 peer-group CORE
 neighbor 195.111.255.6 peer-group CORE
```

```
 neighbor 195.111.255.7 peer-group CORE
 neighbor 195.111.255.8 peer-group CORE
 neighbor 195.111.255.100 peer-group LONDON-RR-CLUSTER
 neighbor 195.111.255.200 peer-group LONDON-RR-CLUSTER
 neighbor 195.111.255.210 peer-group LONDON-RR-CLUSTER
 no auto-summary
!
! Configure the use of Communities in the format X:Y
ip bgp-community new-format
!
route-map our-aggregates permit 10
 match ip address our-aggregates
 set community OurAggregates
!
! Configure the Rendezvous Point using a static anycast address
ip pim rp-address 195.111.255.254
!
! Configure MSDP neighbours on each of the other RPs
ip msdp peer 195.111.255.4 connect-source loopback0
ip msdp peer 195.111.255.6 connect-source loopback0
ip msdp peer 195.111.255.8 connect-source loopback0
!
! Configure the RADIUS servers
aaa group server my-radius-group
 server host 195.111.253.2 auth-port 1812 acct-port 1813 key <string>
 server host 195.111.254.2 auth-port 1812 acct-port 1813 key <string>
!
! Configure the SNMP servers and traps
snmp-server community <string> ro 99
snmp-server host 195.111.253.5 traps version 2c <string>
snmp-server host 195.111.253.6 traps version 2c <string>
snmp-server enable traps
snmp-server trap-source loopback0
snmp-server contact Netops +44 1999 999999
snmp-server location London
!
! Configure NTP servers
ntp bootserver 195.111.253.4
ntp server 195.111.253.4
ntp server 195.111.254.4
!
! Disable HTTP configuration
no ip http server
no ip http server-secure
!
login banner ^C
               +----------------------------------------+
               |                                        |
               | ********* W A R N I N G ********** |
               |                                        |
               |    STRICTLY NO UNAUTHORISED ACCESS     |
               |                                        |
               +----------------------------------------+
^C
```

```
!
line con 0
 ! Use RADIUS to authenticate console access
 login authentication use-radius
 exec-timeout 0 0
 transport preferred none
line vty 0 4
 ! Use RADIUS to authenticate telnet/SSH access
 login authentication use-radius
 exec-timeout 0 0
 transport preferred none
```

12.3 AGGREGATION ROUTER (PE) RUNNING MPLS L3 AND L2VPN OVER LDP, BGP POLICY TO CUSTOMERS, MBGP, PIM-SM (JUNOS)

```
groups {
    re0 {
        system {
            host-name London-Access1-re0;
        }
        interfaces {
            fxp0 {
                unit 0 {
                    family inet {
                        address 10.0.0.101/24;
                    }
                }
            }
        }
    }
    re1 {
        system {
            host-name London-Access1-re1;
        }
        interfaces {
            fxp0 {
                unit 0 {
                    family inet {
                        address 10.0.0.102/24;
                    }
                }
            }
        }
    }
    e1-template {
        interfaces {
```

```
            <e1-*> {
                e1-options {
                    framing g704;
                    fcs 32;
                }
            }
        }
    }
}
apply-groups [ re0 re1 ];
system {
    domain-name exampledomain.com;
    time-zone Europe/London;
    no-redirects;
    dump-on-panic;
    mirror-flash-on-disk;
    backup-router 10.0.0.254 destination 10.0.0.0/8;
    authentication-order radius;
    ports {
        console type vt100;
    }
    archival configuration {
        transfer-interval 1440;
        transfer-on-commit;
        archive-sites {
            ftp://robot-user:robot-pwd@195.111.253.8/data/configs;
        }
    }
    root-authentication {
        encrypted-password "<encrypted password>"; # SECRET-DATA
    }
    name-server {
        195.111.253.1;
        195.111.254.1;
    }
    radius-server {
        195.111.253.2 {
            secret "<encrypted secret key>"; # SECRET-DATA
            timeout 1;
            retry 2;
        }
        195.111.254.2 {
            secret "<encrypted secret key>"; # SECRET-DATA
            timeout 1;
            retry 2;
        }
    }
    login {
        message "\n          +---------------------------------------+ \n      |
|\n        | ********* W A R N I N G ********* |\n            |
|\n        |   STRICTLY NO UNAUTHORISED ACCESS   |\n            |
|\n          +---------------------------------------+\n\n";
        class mysuperuser {
```

```
            idle-timeout 60;
            permissions all;
        }
        class operator {
            permissions [ admin configure interface interface-control network
                routing snmp system trace view firewall ];
            allow-commands "clear interfaces statistics";
        }
        class viewall {
            idle-timeout 60;
            permissions [ admin interface network routing snmp system trace
                view firewall ];
        }
        user mysuperuser {
            full-name "Template RADIUS Account";
            uid 1000;
            class mysuperuser;
        }
        user operator {
            full-name "Template RADIUS Account";
            uid 1001;
            class operator;
        }
        user viewall {
            full-name "Template RADIUS Account";
            uid 1002;
            class viewall;
        }
        user admin {
            full-name "Last resort account";
            uid 9999;
            class mysuperuser;
            authentication {
                encrypted-password "<encrypted password>";
            }
        }
    }
    services {
        inactive: ftp {
            connection-limit 1;
            rate-limit 1;
        }
        ssh {
            root-login deny;
            protocol-version 2;
            connection-limit 4;
            rate-limit 4;
        }
        inactive: telnet {
            connection-limit 4;
            rate-limit 4;
        }

    }
```

```
    syslog {
        archive size 10m files 10;
        user * {
            any emergency;
        }
        host 195.111.253.3 {
            any info;
        }
        host 195.111.254.3 {
            any info;
        }
        file messages {
            any info;
            authorization info;
        }
        file security.log {
            interactive-commands any;
            archive size 10m files 10;
        }
    }.
    ntp {
        boot-server 195.111.253.4;
        server 195.111.253.4;
        server 195.111.254.4;
    }
}
chassis {
    no-reset-on-timeout;
    dump-on-panic;
    redundancy {
        routing-engine 0 master;
        routing-engine 1 backup;
        failover on-loss-of-keepalives;
        keepalive-time 300;
    }
    fpc 0 {
        pic 0 {
            framing sdh;
            vtmapping klm;
        }
    }
}
interfaces {
    apply-groups e1-template;
    e1-0/0/0:0 {
        description "E1 to customer A";
        encapsulation ppp;
        unit 0 {
            family inet {
                address 158.160.0.1/30;
            }
        }
    }
    e1-0/0/0:1 {
        description "E1 to customer B (two sites via Frame Relay)";
```

```
        encapsulation frame-relay;
        unit 100 {
            dlci 100;
            dce;
            family inet {
                address 158.160.0.5/30 {
                    destination 158.160.0.6;
                }
            }
        }
        unit 101 {
            dlci 101;
            dce;
            family inet {
                address 158.160.0.9/30 {
                    destination 158.160.0.10;
                }
            }
        }
        unit 200 {
            dlci 200;
            dce;
        }
        unit 201 {
            dlci 201;
            dce;
        }
    }
    e1-0/0/0:2 {
        description "E1 to customer C";
        encapsulation cisco-hdlc;
        unit 0 {
            family inet {
                address 158.160.0.13/30;
            }
        }
    }
    at-0/1/0 {
        description "ATM customer links";
        encapsulation atm-pvc;
        atm-options {
            pic-type atm1;
            vpi 0 max-vcs 1024;
        }
        unit 200 {
            description "ATM PVC to customer D";
            vci 0.200;
            encapsulation atm-snap;
            family inet {
                address 158.160.1.1/30 {
                    destination 158.160.1.2;
                }
            }
        }
```

```
    unit 201 {
        description "ATM PVC to customer E";
        vci 0.201;
        encapsulation atm-snap;
        family inet {
            address 158.160.1.5/30 {
                destination 158.160.1.6;
            }
        }
    }
}
atm-0/2/0 {
    description "More ATM customer links";
    encapsulation atm-pvc;
    atm-options {
        pic-type atm2;
        vpi 0 {
            shaping {
                cbr peak 100m sustained 10m burst 50;
            }
        }
        vpi 1 {
            shaping {
                vbr peak 100m sustained 5m burst 25;
            }
        }
    }
    unit 101 {
        description "ATM PVC to customer F";
        vci 0.101;
        encapsulation atm-snap;
        family inet {
            address 158.160.2.1/30 {
                destination 158.160.2.2;
            }
        }
    }
    unit 1101 {
        description "ATM PVC to customer G";
        vci 1.101;
        encapsulation atm-snap;
        family inet {
            address 158.160.2.5/30 {
                destination 158.160.2.6;
            }
        }
    }
}
ge-2/0/0 {
    description "GigE to primary local LAN";
    unit 0 {
        family inet {
            address 195.111.100.101/24;
        }
```

```
                family iso;
                family mpls;
            }
        }
    ge-3/0/0 {
        description "GigE to backup local LAN";
        unit 0 {
            family inet {
                address 195.111.101.101/24;
            }
            family iso;
            family mpls;
        }
    }
    ge-3/0/1 {
        description "GigE to customer H";
        vlan-tagging;
        unit 100 {
            vlan-id 100;
            family inet {
                address 158.160.8.1/30;
            }
        }
    }
    lo0 {
        unit 0 {
            family inet {
                filter {
                    input accessfilter;
                }
                address 195.111.255.101/32;
            }
            family iso {
                address 49.0000.1921.6825.5101.00;
            }
        }
    }
}
forwarding-options {
    sampling {
        input {
            family inet {
                rate 500;
            }
        }
        output {
            cflowd 195.111.254.5 {
                version 5;
                local-dump;
                autonomous-system-type origin;
            }
        }
    }
}
```

```
snmp {
    description "London-PE1 Juniper M40e";
    location "London";
    contact "Netops +44 1999 999999";
    community <string> {
        authorization read-only;
        clients {
            195.111.253.5/32;
            195.111.253.6/32;
        }
    }
    trap-options {
        source-address lo0;
    }
    trap-group Juniper {
        version v2;
        categories {
            authentication;
            chassis;
            link;
            routing;
            startup;
        }
        targets {
            195.111.253.5;
            195.111.253.6;
        }
    }
}
routing-options {
    static {
        route 10.0.0.0/8 {
            next-hop 10.0.0.254;
            retain;
            no-readvertise;
        }
    }
    autonomous-system 1;
    forwarding-table {
        export load-share;
        unicast-reverse-path feasible-paths;
    }
    interface-routes {
        rib-group inet interface-rib;
    }
    rib-groups {
        bgp-rib {
            import-rib [ inet.0 inet.2 ];
            export-rib [ inet.0 ];
        }
        multicast-rib {
            import-rib [ inet.2 ];
            export-rib [ inet.2 ];
        }
```

```
        interface-rib {
            import-rib [ inet.0 inet.2 ];
        }
    }
    multicast {
        scope customerA-local {
            interface e1-0/0/0:0.0;
            prefix 239.255.0.0/24;
        }
    }
}
protocols {
    mpls {
        traceoptions {
            file mpls.log files 10 size 1m;
            flag hello detail;
            flag state detail;
        }
        log-updown;
        no-propagate-ttl;
        interface ge-2/0/0.0;
        interface ge-3/0/0.0;
        interface lo0.0;
        interface fxp0.0 {
            disable;
        }
    }
    ldp {
        traceoptions {
            file ldp.log files 10 size 1m;
            flag hello detail;
            flag state detail;
        }
        log-updown;
        interface ge-2/0/0.0;
        interface ge-3/0/0.0;
        interface lo0.0;
        interface fxp0.0 {
            disable;
        }
    }
    bgp {
        traceoptions {
            file bgp.log files 10 size 1m;
            flag open detail;
            flag state detail;
        }
        rib-group bgp-rib;
        log-updown;
        group LONDON-RR-CLUSTER {
            type internal;
            export next-hop-self;
            local-address 195.111.255.101;
            neighbor 195.111.255.1;
```

```
            neighbor 195.111.255.2;
            family inet {
                any;
            }
            family l2vpn {
                unicast;
            }
            family l3vpn {
                any;
            }
        }
    group CUSTOMERS {
        type external;
        export to-all-customers;
        neighbor 158.160.0.2 {
            description "Customer A";
            peer-as 1000;
            import from-customerA;
        }
        neighbor 158.160.0.6 {
            description "Customer B, Site 1";
            peer-as 1001;
            import from-customerB;
        }
        neighbor 158.160.0.10 {
            description "Customer B, Site 2";
            peer-as 1001;
            import from-customerB;
        }
        neighbor 158.160.0.14 {
            description "Customer C";
            peer-as 1002;
            import from-customerC;
        }
        neighbor 158.160.1.2 {
            description "Customer D";
            peer-as 1003;
            import from-customerD;
        }
        neighbor 158.160.1.6 {
            description "Customer E";
            peer-as 1004
            import from-customerE;
        }
        neighbor 158.160.2.2 {
            description "Customer F";
            peer-as 1005;
            import from-customerF;
        }
        neighbor 158.160.2.6 {
            description "Customer G";
            peer-as 1006;
            import from-customerG;
        }
```

```
        }
    }
    isis {
        traceoptions {
            file isis.log files 10 size 1m;
            flag hello detail;
            flag error detail;
            flag state detail;
        }
        multicast-topology;
        interface ge-2/0/0.0 {
            level 1 {
                metric 25;
                hello-interval 10;
                hold-time 40;
            }
            level 2 disable;
        }
        interface ge-3/0/0.0 {
            level 1 {
                metric 30;
                hello-interval 10;
                hold-time 40;
            }
            level 2 disable;
        }
        interface lo0.0;
        interface fxp0.0 {
            disable;
        }
    }
    pim {
        traceoptions {
            file pim.log files 10 size 1m;
            flag hello detail;
            flag join detail;
            flag rp detail;
            flag assert detail;
            flag state detail;
        }
        rib-group multicast-rib;
        rp {
            static {
                address 195.111.255.254 {
                    group-ranges 224.0.0.0/4;
                }
            }
        }
        interface e1-0/0/0:0.0 {
            mode sparse;
        }
        interface e1-0/0/0:1.100 {
            mode sparse;
        }
```

```
        interface e1-0/0/0:1.101 {
            mode sparse;
        }
        interface e1-0/0/0:2.0 {
            mode sparse;
        }
        interface at-0/0/1.200 {
            mode sparse;
        }
        interface at-0/1/0.201 {
            mode sparse;
        }
        interface at-0/2/0.101 {
            mode sparse;
        }
        interface at-0/2/0.1101 {
            mode sparse;
        }
        interface ge-2/0/0.0{
            mode sparse;
        }
        interface ge-3/0/0.0{
            mode sparse;
        }
        interface lo0.0{
            mode sparse;
        }
        interface fxp0.0 {
            disable;
        }
    }
}
policy-options {
    prefix-list my-martians {
        0.0.0.0/7;
        2.0.0.0/8;
        5.0.0.0/8;
        7.0.0.0/8;
        10.0.0.0/8;
        23.0.0.0/8;
        27.0.0.0/8;
        31.0.0.0/8;
        36.0.0.0/7;
        39.0.0.0/8;
        41.0.0.0/8;
        42.0.0.0/8;
        49.0.0.0/8;
        50.0.0.0/8;
        58.0.0.0/7;
        70.0.0.0/7;
        72.0.0.0/5;
        83.0.0.0/8;
        84.0.0.0/6;
        88.0.0.0/5;
```

```
        96.0.0.0/3;
        169.254.0.0/16;
        172.16.0.0/12;
        173.0.0.0/8;
        174.0.0.0/7;
        176.0.0.0/5;
        184.0.0.0/6;
        189.0.0.0/8;
        190.0.0.0/8;
        192.0.2.0/24;
        192.168.0.0/16;
        197.0.0.0/8;
        198.18.0.0/15;
        223.0.0.0/8;
        224.0.0.0/3;
    }
    prefix-list our-aggregates {
        195.111.0.0/16;
    }
    prefix-list customer-A {
        <customer-A's prefixes>;
    }
    prefix-list customer-B {
        <customer-B's prefixes>;
    }
    prefix-list customer-C {
        <customer-C's prefixes>;
    }
    prefix-list customer-D {
        <customer-D's prefixes>;
    }
    prefix-list customer-E {
        <customer-E's prefixes>;
    }
    prefix-list customer-F {
        <customer-F's prefixes>;
    }
    prefix-list customer-G {
        <customer-G's prefixes>;
    }
    policy-statement load-share {
        term load-balance-per-flow {
            then {
                load-balance per-packet;
            }
        }
    }
    policy-statement next-hop-self {
        term set-next-hop-self {
            from protocol bgp;
            then next-hop-self;
        }
    }
    policy-statement transit-in {
```

```
        term drop-illegal-prefixes {
            from {
                protocol bgp;
                prefix-list my-martians;
            }
            then reject;
        }
        term accept-and-mark {
            from protocol bgp;
            then {
                community set Transit;
                accept;
            }
        }
    }
    policy-statement transit-out {
        term block-peer-routes {
            from {
                protocol bgp;
                community Peer;
            }
            then reject;
        }
        term block-transit-routes {
            from {
                protocol bgp;
                community Transit;
            }
            then reject;
        }
        term send-our-aggregates {
            from {
                protocol bgp;
                community OurAggregates;
            }
            then accept;
        }
        term leak-dual-homed-customer-routes {
            from {
                protocol bgp;
                community PleaseLeakThis;
            }
            then accept;
        }
        term block-more-specifics {
            from {
                protocol bgp;
                prefix-list our-aggregates;
            }
            then reject;
        }
        term send-customer-routes {
            from {
                protocol bgp;
```

```
                        community Customer;
                }
            then accept;
        }
    }
    policy-statement to-all-customers {
        term PleaseLeakThis {
            from {
                protocol bgp;
                community PleaseLeakThis;
            }
            then accept;
        }
        term DontLeakLongerPrefixes {
            from {
                protocol bgp;
                prefix-list our-aggregates;
            }
            then reject;
        }
        term CustomerRoutes {
            from {
                protocol bgp;
                community Customer;
            }
            then accept;
        }
        term PeerRoutes {
            from {
                protocol bgp;
                community Peer;
            }
            then accept;
        }
        term TransitRoutes {
            from {
                protocol bgp;
                community Transit;
            }
            then accept;
        }
        then reject;
    }
    policy-statement from-customer-A {
        term customer-A-prefixes {
            from {
                protocol bgp;
                prefix-list customer-A;
            }
            then accept;
        }
        then reject;
    }
    policy-statement from-customer-B {
```

```
        term customer-B-prefixes {
            from {
                protocol bgp;
                prefix-list customer-B;
            }
            then accept;
        }
        then reject;
    }
    policy-statement from-customer-C {
        term customer-C-prefixes {
            from {
                protocol bgp;
                prefix-list customer-C;
            }
            then accept;
        }
        then reject;
    }
    policy-statement from-customer-D {
        term customer-D-prefixes {
            from {
                protocol bgp;
                prefix-list customer-D;
            }
            then accept;
        }
        then reject;
    }
    policy-statement from-customer-E {
        term customer-E-prefixes {
            from {
                protocol bgp;
                prefix-list customer-E;
            }
            then accept;
        }
        then reject;
    }
    policy-statement from-customer-F {
        term customer-F-prefixes {
            from {
                protocol bgp;
                prefix-list customer-F;
            }
            then accept;
        }
        then reject;
    }
    policy-statement from-customer-G {
        term customer-G-prefixes {
            from {
                protocol bgp;
                prefix-list customer-G;
```

```
                }
                then accept;
            }
            then reject;
        }
        policy-statement vpn-custB-in {
            from {
                protocol bgp;
                community vpn-custB;
            }
            then accept;
        }
        policy-statement vpn-custB-out {
            from protocol bgp;
            then {
                community add vpn-custB;
                accept;
            }
        }
        community Customer members 1:1;
        community Peer members 1:2;
        community Transit members 1:3;
        community DontLeakThis members 1:999;
        community PleaseLeakThis members 1:1000;
        community vpn-custA members 1:5001;
        community vpn-custB members 1:5002;
        community vpn-custH members 1:5003;
        community OurAggregates 1:50000;
    }
    routing-instances {
        customerA-vpn {
            instance-type l3vpn;
            route-distinguisher 1:1;
            interface e1-0/0/0:0.0;
            vrf-target vpn-custA;
            protocols bgp {
                type external;
                peer-as 64512;
                neighbor 158.160.0.2;
            }
            routing-options {
                autonomous-system 1 loops 1;
                maximum-routes 10;
            }
        }
        customerB-vpn {
            instance-type l2vpn;
            route-distinguisher 1:2;
            interface e1-0/0/0:1.200;
            interface e1-0/0/0:1.201;
            vrf-import vpn-custB-in;
            vrf-export vpn-custB-out;
            protocols {
                l2vpn {
```

```
                    encapsulation-type frame-relay;
                    site site1 {
                        site-identifier 1;
                        interface e1-0/0/0:1.200 {
                            remote-site-id 2;
                        }
                        interface e1-0/0/0:1.201 {
                            remote-site-id 3;
                        }
                    }
                }
            }
        }
    customerH-vpn {
        instance-type l3vpn;
        route-distinguisher 1:3;
        interface ge-3/0/1.100;
        vrf-target vpn-custH;
        vrf-table-label;
        protocols bgp {
            type external;
            peer-as 64513;
            neighbor 158.60.8.2;
        }
        routing-options {
            autonomous-system 1 loops 1;
            maximum-routes 25;
        }
    }
}
class-of-service {
    classifiers {
        exp mpls-exp {
            forwarding-class expedited-forwarding {
                loss-priority high code-points [ 101 110 111 ];
            }
            forwarding-class assured-forwarding {
                loss-priority high code-points 011;
            }
            forwarding-class best-effort {
                loss-priority high code-points [ 000 010 100 ];
            }
            forwarding-class less-than-best-effort {
                loss-priority high code-points 001;
            }
        }
        dscp diffserv {
            forwarding-class expedited-forwarding {
                loss-priority high code-points [ ef cs6 cs7 cs5 ];
            }
            forwarding-class assured-forwarding {
                loss-priority high code-points af31;
                loss-priority low code-points [ af32 af33 ];
            }
```

```
                forwarding-class best-effort {
                    loss-priority high code-points [ be cs2 cs4 ];
                }
                forwarding-class less-than-best-effort {
                    loss-priority high code-points cs1;
                }
            }
            inet-precedence inet-prec {
                forwarding-class expedited-forwarding {
                    loss-priority high code-points [ 101 110 111 ];
                }
                forwarding-class assured-forwarding {
                    loss-priority high code-points 011;
                }
                forwarding-class best-effort {
                    loss-priority high code-points [ 000 010 100 ];
                }
                forwarding-class less-than-best-effort {
                    loss-priority high code-points 001;
                }
            }
        }
        drop-profiles {
            aggressive {
                fill-level 10 drop-probability 30;
                fill-level 70 drop-probability 100;
            }
            gentle {
                fill-level 20 drop-probability 20;
                fill-level 80 drop-probability 80;
            }
        }
        forwarding-classes {
            queue 0 best-effort;
            queue 1 less-than-best-effort;
            queue 2 assured-forwarding;
            queue 3 expedited-forwarding;
        }
        interfaces {
            e1-0/0/0:* {
                scheduler-map diffserv;
                unit * {
                    classifiers {
                        inet-precedence inet-prec;
                    }
                }
            }
            at-0/1/0 {
                scheduler-map diffserv;
                unit * {
                    classifiers {
                        inet-precedence inet-prec;
                    }
                }
```

```
        }
    at-0/2/0 {
        scheduler-map diffserv;
        unit * {
            classifiers {
                inet-precedence inet-prec;
            }
        }
    }
    ge-2/*/* {
        scheduler-map diffserv;
        unit * {
            classifiers {
                dscp diffserv;
                exp mpls-exp;
            }
        }
    }
    ge-3/*/* {
        scheduler-map diffserv;
        unit * {
            classifiers {
                dscp diffserv;
                exp mpls-exp;
            }
        }
    }
}
scheduler-maps {
    diffserv {
        forwarding-class expedited-forwarding scheduler
            expedited-forwarding;
        forwarding-class assured-forwarding scheduler assured-forwarding;
        forwarding-class best-effort scheduler best-effort;
        forwarding-class less-than-best-effort scheduler
            less-than-best-effort;
    }
}
schedulers {
    expedited-forwarding {
        transmit-rate percent 10 exact;
        buffer-size percent 10;
        priority strict-high;
    }
    assured-forwarding {
        transmit-rate percent 10;
        buffer-size percent 10;
        priority high;
        drop-profile-map loss-priority high protocol any drop-profile
            aggressive;
        drop-profile-map loss-priority low protocol any drop-profile
            gentle;
    }
    best-effort {
```

```
                transmit-rate percent 79;
                buffer-size percent 79;
                priority low;
                drop-profile-map loss-priority any protocol any drop-profile
                    gentle;
            }
            less-than-best-effort {
                transmit-rate remainder;
                buffer-size percent 1;
                priority low;
                drop-profile-map loss-priority any protocol any drop-profile
                    gentle;
            }
        }
    }
}
firewall {
    filter accessfilter {
        term allow-telnet-from-very-small-list {
            from {
                source-address {
                    195.111.200.0/24;
                    195.111.255.0/24;
                }
                protocol tcp;
                destination-port telnet;
            }
            then accept;
        }
        term deny-telnet-from-everywhere-else {
            from {
                protocol tcp;
                destination-port telnet;
            }
            then {
                log;
                reject;
            }
        }
        term allow-ftp-from-very-small-list {
            from {
                source-address {
                    195.111.200.0/24;
                }
                protocol tcp;
                port [ ftp ftp-data ];
            }
            then accept;
        }
        term deny-ftp-from-everywhere-else {
            from {
                protocol tcp;
                destination-port [ ftp ftp-data ];
            }
            then {
```

```
                        log;
                        reject;
                    }
                }
            term allow-everything-else {
                then accept;
            }
        }
    filter cflowd {
        term sample-and-accept {
            then {
                sample;
                accept;
            }
        }
    }
}
```

12.4   AGGREGATION ROUTER (PE) RUNNING MPLS L3 AND L2VPN OVER LDP, BGP POLICY TO CUSTOMERS, MBGP, PIM-SM (IOS)

```
no service tcp-small-servers
no service udp-small-servers
service password-encryption
no service pad
no ip finger
no ip bootp server
!
aaa new-model
aaa authentication login use-radius group radius local
aaa authorization exec default group radius
aaa accounting commands 0 default start-stop group radius
aaa accounting system default start-stop group radius
aaa accounting connection default start-stop group radius
!
enable secret 5 XXXXXXXXX
!
username admin password 7 XXXXXXXXX
hostname London-Access1
ip domain-name exampledomain.com
ip name-server 195.111.253.1 195.111.254.1
!
ip ftp username robot-user
ip ftp password 0 robot-pwd
exception protocol ftp
exception dump 195.111.253.8
```

```
exception core-file London-Access1-corefile
!
ip cef distributed
!
mpls ip
mpls ldp advertise-labels for 1
mpls ldp router-id lo0
mpls ldp discovery targeted-hellos accept
!
ip multicast-routing
!
logging 195.111.253.3 6
logging 195.111.254.3 6
logging console 0
logging source-interface lo0
!
ip flow-export 195.111.254.5 9995 version 5 origin-as
!
interface Serial0/0/0:0
 description E1 to Customer A
 encapsulation ppp
 ip address 158.160.0.1 255.255.255.252
 ip access-group accessfilter in
 service-policy output core-forwarding
 ip vrf forwarding customerA-vpn
 ip pim sparse-mode
!
interface Serial0/0/0:1
 description E1 to Customer B (Frame Relay with two sites)
 encapsulation frame-relay ietf
!
interface Serial0/0/0:1.100 point-to-point
 description Frame Relay PVC to customer B site 1
 encapsulation frame-relay ietf
 frame-relay interface-dlci 100
 ip address 158.160.0.5 255.255.255.252
 ip pim sparse-mode
!
interface Serial0/0/0:1.101 point-to-point
 description Frame Relay PVC to customer B site 2
 encapsulation frame-relay ietf
 frame-relay interface-dlci 101
 ip address 158.160.0.9 255.255.255.252
 ip pim sparse-mode
!
interface Serial0/0/0:1.200 point-to-point
 description Frame Relay PVC to customer B l2vpn 1
 encapsulation frame-relay ietf
 frame-relay interface-dlci 200
 xconnect 195.111.255.131 encapsulation mpls pw-class customerB-vpn
!
interface Serial0/0/0:1.201 point-to-point
 description Frame Relay PVC to customer B l2vpn 2
 encapsulation frame-relay ietf
```

```
frame-relay interface-dlci 201
xconnect 195.111.255.142 encapsulation mpls pw-class customerB-vpn
!
interface Serial0/0/0:2
 description E1 to Customer C
 encapsulation hdlc
 ip address 158.160.0.13 255.255.255.252
 ip pim sparse-mode
!
interface ATM0/1/0
 description ATM interface to various customers
!
interface ATM0/1/0.201
 description ATM PVC to customer D
 atm pvc 201 0 201 aal5snap
 ip address 158.160.1.1 255.255.255.252
 ip pim sparse-mode
!
interface ATM0/1/0.202
 description ATM PVC to customer E
 atm pvc 202 0 202 aal5snap
 ip address 158.160.1.5 255.255.255.252
 ip pim sparse-mode
!
interface ATM0/2/0
 description More ATM links to customers
!
interface ATM0/2/0.101
 description ATM PVC to customer F
 atm pvc 101 0 101 aal5snap 100000 10000 50
 ip address 158.160.2.1 255.255.255.252
 ip pim sparse-mode
!
interface ATM0/2/0.1101
 description ATM PVC to customer G
 atm pvc 1101 1 101 aal5snap 100000 5000 25
 ip address 158.160.2.5 255.255.255.252
 ip pim sparse-mode
!
interface GigabitEthernet2/0
 description GigE to primary local LAN
 ip address 195.111.100.101 255.255.255.0
 ip access-group accessfilter in
 service-policy output core-forwarding
 ip router isis
 isis circuit-type level-1
 isis metric 25 level-1
 isis hello-interval 10 level-1
 isis hello-multiplier 4
 mpls ip
 mpls label protocol ldp
 ip pim sparse-mode
 !
interface GigabitEthernet3/0
```

```
  description GigE to secondary local LAN
  ip address 195.111.101.101 255.255.255.0
  ip access-group accessfilter in
  service-policy output core-forwarding
  ip router isis
  isis circuit-type level-1
  isis metric 30 level-1
  isis hello-interval 10 level-1
  isis hello-multiplier 4
  mpls ip
  mpls label protocol ldp
  ip pim sparse-mode
!
interface GigabitEthernet3/1
  description GigE to customer H
  ip address 158.160.8.1 255.255.255.0
  service-policy output core-forwarding
  ip vrf forwarding customerH-vpn
  ip pim sparse-mode
!
interface Ethernet 4/0/0
  description Link to Management Network
  ip address 10.0.0.101 255.255.255.0
  ip access-group mgmt-in in
  ip access-group mgmt-out out
!
interface lo0
  description Loopback
  ip address 195.111.255.101 255.255.255.255
  ip router isis
!
interface loopback1
  description Loopback for CustomerA-vpn
  ip address 192.168.0.1 255.255.255.255
  ip vrf forwarding customerA-vpn
!
interface loopback8
  description Loopback for CustomerH-vpn
  ip address 192.168.0.8 255.255.255.255
  ip vrf forwarding customerH-vpn
!
ip vrf customerA-vpn
  rd 1:1
  route-target import 1:5001
  route-target export 1:5001
!
ip vrf customerH-vpn
  rd 1:3
  route-target import 1:5003
  route-target export 1:5003
!
pseudowire-class customerB-vpn
  encapsulation mpls
  interworking ip
```

```
!
ip access-list 1 permit 195.111.255.0 0.0.0.255
ip access-list 99 permit host 195.111.253.5
ip access-list 99 permit host 195.111.253.6
!
ip access-list extended mgmt-in
 permit 10.0.0.0 0.0.0.255 host 195.111.255.1
 permit 10.0.0.0 0.0.0.255 host 10.0.0.1
 deny any any
ip access-list extended mgmt-out
 permit host 195.111.255.1 10.0.0.0 0.255.255.255
 permit host 10.0.0.1 10.0.0.0 0.255.255.255
 deny any any
ip access-list extended accessfilter
 permit ip 195.111.250.0 0.0.0.255 host 195.111.255.1 eq ssh
 permit ip 195.111.250.0 0.0.0.255 195.111.0.0 0.0.3.255 eq ssh
 permit ip 195.111.255.0 0.0.0.255 host 195.111.255.1 eq ssh
 permit ip 195.111.255.0 0.0.0.255 195.111.0.0 0.0.3.255 eq ssh
 permit ip 10.24.0.0 0.0.0.255 host 195.111.255.1 eq ssh
 permit ip 10.24.0.0 0.0.0.255 195.111.0.0 0.0.3.255 eq ssh
 deny ip any host 195.111.255.1 eq ssh
 deny ip any 195.111.0.0 0.0.3.255 eq ssh
 permit ip 195.111.250.0 0.0.0.255 host 195.111.255.1 eq ftp
 permit ip 195.111.250.0 0.0.0.255 host 195.111.255.1 eq ftp-data
 permit ip 195.111.255.0 0.0.0.255 host 195.111.255.1 eq ftp
 permit ip 195.111.255.0 0.0.0.255 host 195.111.255.1 eq ftp-data
 permit ip 195.111.250.0 0.0.0.255 195.111.0.0 0.0.3.255 eq ftp
 permit ip 195.111.250.0 0.0.0.255 195.111.0.0 0.0.3.255 eq ftp-data
 permit ip 195.111.255.0 0.0.0.255 195.111.0.0 0.0.3.255 eq ftp
 permit ip 195.111.255.0 0.0.0.255 195.111.0.0 0.0.3.255 eq ftp-data
 deny ip any host 195.111.255.1 eq ftp
 deny ip any host 195.111.255.1 eq ftp-data
 deny ip any 195.111.0.0 0.0.3.255 eq ftp
 deny ip any 195.111.0.0 0.0.3.255 eq ftp
 permit ip 195.111.250.0 0.0.0.255 host 195.111.255.1 eq telnet
 permit ip 195.111.250.0 0.0.0.255 195.111.0.0 0.0.3.255 eq telnet
 permit ip 195.111.255.0 0.0.0.255 host 195.111.255.1 eq telnet
 permit ip 195.111.255.0 0.0.0.255 195.111.0.0 0.0.3.255 eq telnet
 deny ip any host 195.111.255.1 eq ssh
 deny ip any 195.111.0.0 0.0.3.255 eq ssh
 permit ip any any
!
ip access-list standard customer-A
 permit <customer-A's prefixes>
!
ip access-list standard customer-B
 permit <customer-B's prefixes>
!
ip access-list standard customer-C
 permit <customer-C's prefixes>
!
ip access-list standard customer-D
 permit <customer-D's prefixes>
!
```

```
ip access-list standard customer-E
 permit <customer-E's prefixes>
!
ip access-list standard customer-F
 permit <customer-F's prefixes>
!
ip access-list standard customer-G
 permit <customer-G's prefixes>
!
ip prefix-list my-martians deny 0.0.0.0/7
ip prefix-list my-martians deny 2.0.0.0/8
ip prefix-list my-martians deny 5.0.0.0/8
ip prefix-list my-martians deny 7.0.0.0/8
ip prefix-list my-martians deny 10.0.0.0/8
ip prefix-list my-martians deny 23.0.0.0/8
ip prefix-list my-martians deny 27.0.0.0/8
ip prefix-list my-martians deny 31.0.0.0/8
ip prefix-list my-martians deny 36.0.0.0/7
ip prefix-list my-martians deny 39.0.0.0/8
ip prefix-list my-martians deny 41.0.0.0/8
ip prefix-list my-martians deny 42.0.0.0/8
ip prefix-list my-martians deny 49.0.0.0/8
ip prefix-list my-martians deny 50.0.0.0/8
ip prefix-list my-martians deny 58.0.0.0/7
ip prefix-list my-martians deny 70.0.0.0/7
ip prefix-list my-martians deny 72.0.0.0/5
ip prefix-list my-martians deny 83.0.0.0/8
ip prefix-list my-martians deny 84.0.0.0/6
ip prefix-list my-martians deny 88.0.0.0/5
ip prefix-list my-martians deny 96.0.0.0/3
ip prefix-list my-martians deny 169.254.0.0/16
ip prefix-list my-martians deny 172.16.0.0/12
ip prefix-list my-martians deny 173.0.0.0/8
ip prefix-list my-martians deny 174.0.0.0/7
ip prefix-list my-martians deny 176.0.0.0/5
ip prefix-list my-martians deny 184.0.0.0/6
ip prefix-list my-martians deny 189.0.0.0/8
ip prefix-list my-martians deny 190.0.0.0/8
ip prefix-list my-martians deny 192.0.2.0/24
ip prefix-list my-martians deny 192.168.0.0/16
ip prefix-list my-martians deny 197.0.0.0/8
ip prefix-list my-martians deny 198.18.0.0/15
ip prefix-list my-martians deny 223.0.0.0/8
ip prefix-list my-martians deny 224.0.0.0/3
ip prefix-list my-martians permit 0.0.0.0/0 ge 8 le 28
!
ip community-list standard Customer permit 1:1
ip community-list standard Peer permit 1:2
ip community-list standard Transit permit 1:3
ip community-list standard DontLeakThis permit 1:999
ip community-list standard PleaseLeakThis permit 1:1000
ip community-list standard OurAggregates permit 1:50000
!
class-map match-any expedited-forwarding
```

```
 match ip precedence 101 110 111
 match mpls experimental 101 110 111
 match ip dscp ef cs5 cs6 cs7
class-map match-any assured-forwarding
 match ip precedence 011
 match mpls experimental 011
 match ip dscp af31 af32 af33
class-map match-any best-effort
 match ip precedence 000 010 100
 match mpls experimental 000 010 100
 match ip dscp default cs2 cs4
class-map match-any less-than-best-effort
 match ip precedence 001
 match mpls experimental 001
 match ip dscp cs1
!
policy-map core-forwarding
 class expedited-forwarding
 bandwidth percent 10
 class assured-forwarding
 bandwidth percent 10
 class best-effort
 bandwidth percent 79
 class less-than-best-effort
 bandwidth percent 1
!
router isis
 net 49.0000.1921.6825.5101.00
 is-type level-2
!
router bgp 1
 no synchronize
 neighbor LONDON-RR-CLUSTER peer-group
 neighbor LONDON-RR-CLUSTER remote-as 1
 neighbor LONDON-RR-CLUSTER send-community
 neighbor LONDON-RR-CLUSTER update-source loopback0
 neighbor ALL-CUSTOMERS peer-group
 neighbor ALL-CUSTOMERS send-community
 neighbor ALL-CUSTOMERS update-source loopback0
 neighbor ALL-CUSTOMERS route-map to-all-customers out
 neighbor 195.111.255.1 peer-group LONDON-RR-CLUSTER
 neighbor 195.111.255.2 peer-group LONDON-RR-CLUSTER
 neighbor 158.160.0.2 peer-group ALL-CUSTOMERS
 neighbor 158.160.0.2 remote-as 1000
 neighbor 158.160.0.2 route-map from-customer-A in
 neighbor 158.160.0.2 peer-group ALL-CUSTOMERS
 neighbor 158.160.0.6 remote-as 1001
 neighbor 158.160.0.6 route-map from-customer-B in
 neighbor 158.160.0.2 peer-group ALL-CUSTOMERS
 neighbor 158.160.0.10 remote-as 1001
 neighbor 158.160.0.10 route-map from-customer-B in
 neighbor 158.160.0.2 peer-group ALL-CUSTOMERS
 neighbor 158.160.0.14 remote-as 1002
 neighbor 158.160.0.14 route-map from-customer-C in
```

```
 neighbor 158.160.0.2 peer-group ALL-CUSTOMERS
 neighbor 158.160.1.2 remote-as 1003
 neighbor 158.160.1.2 route-map from-customer-D in
 neighbor 158.160.0.2 peer-group ALL-CUSTOMERS
 neighbor 158.160.1.6 remote-as 1004
 neighbor 158.160.1.6 route-map from-customer-E in
 neighbor 158.160.0.2 peer-group ALL-CUSTOMERS
 neighbor 158.160.2.2 remote-as 1005
 neighbor 158.160.2.2 route-map from-customer-F in
 neighbor 158.160.0.2 peer-group ALL-CUSTOMERS
 neighbor 158.160.2.6 remote-as 1006
 neighbor 158.160.2.6 route-map from-customer-G in
 no auto-summary
 address-family vpnv4
  neighbor 195.111.255.1 activate
  neighbor 195.111.255.1 send-community both
  neighbor 195.111.255.2 activate
  neighbor 195.111.255.2 send-community both
 exit address-family
 address-family ipv4 vrf customerA-vpn
  redistribute connected
  no auto-summary
  no synchronization
 exit address-family
 address-family ipv4 vrf customerH-vpn
  redistribute connected
  no auto-summary
  no synchronization
 exit address-family
!
ip bgp-community new-format
!
route-map to-all-customers permit 5
 match community PleaseLeakThis
!
route-map to-all-customers permit 10
 match community Customer
!
route-map to-all-customers permit 20
 match community Peer
!
route-map to-all-customers permit 30
 match community Transit
!
route-map to-all-customers permit 40
 match community OurAggregates
!
route-map from-customer-A permit 10
 match ip address customer-A
 set community 1:1 additive
!
route-map from-customer-B permit 10
 match ip address customer-B
 set community 1:1 additive
```

```
!
route-map from-customer-C permit 10
 match ip address customer-C
 set community 1:1 additive
!
route-map from-customer-D permit 10
 match ip address customer-D
 set community 1:1 additive
!
route-map from-customer-E permit 10
 match ip address customer-E
 set community 1:1 additive
!
route-map from-customer-F permit 10
 match ip address customer-F
 set community 1:1 additive
!
route-map from-customer-G permit 10
 match ip address customer-G
 set community 1:1 additive
!
ip pim rp-address 195.111.255.254
!
aaa group server my-radius-group
 server host 195.111.253.2 auth-port 1812 acct-port 1813 key <string>
 server host 195.111.254.2 auth-port 1812 acct-port 1813 key <string>
!
snmp-server community <string> ro 99
snmp-server host 195.111.253.5 traps version 2c <string>
snmp-server host 195.111.253.6 traps version 2c <string>
snmp-server enable traps
snmp-server trap-source loopback0
snmp-server contact Netops +44 1999 999999
snmp-server location London
!
ntp bootserver 195.111.253.4
ntp server 195.111.253.4
ntp server 195.111.254.4
!
no ip http server
no ip http server-secure
!
login banner ^C
            +---------------------------------------+
            |                                       |
            | ********** W A R N I N G **********    |
            |                                       |
            |    STRICTLY NO UNAUTHORISED ACCESS    |
            |                                       |
            +---------------------------------------+
^C
!
line con 0
 login authentication use-radius
```

```
exec-timeout 0 0
transport preferred none
line vty 0 4
login authentication use-radius
exec-timeout 0 0
transport preferred none
```

12.5 BORDER ROUTER RUNNING MPLS WITH LDP, BGP POLICY TO PEERS, MBGP, PIM-SM (JUNOS)

```
groups {
    re0 {
        system {
            host-name London-Border1-re0;
        }
        interface {
            fxp0 {
                unit 0 {
                    family inet {
                        address 10.0.0.201/24;
                    }
                }
            }
        }
    }
    re1 {
        system {
            host-name London-Border1-re1;
        }
        interface {
            fxp0 {
                unit 0 {
                    family inet {
                        address 10.0.0.202/24;
                    }
                }
            }
        }
    }
    sdh-template {
        interfaces {
            <so-*/*/*> {
                sonet-options {
                    rfc-2615;
                }
            }
        }
    }
```

```
}
apply-groups [ re0 re1 ];
system {
    domain-name exampledomain.com;
    time-zone Europe/London;
    no-redirects;
    dump-on-panic;
    mirror-flash-on-disk;
    backup-router 10.0.0.254 destination 10.0.0.0/8;
    authentication-order radius;
    ports {
        console type vt100;
    }
    archival configuration {
        transfer-interval 1440;
        transfer-on-commit;
        archive-sites {
            ftp://robot-user:robot-pwd@195.111.253.8/data/configs;
        }
    }
    root-authentication {
        encrypted-password "<encrypted password>"; # SECRET-DATA
    }
    name-server {
        195.111.253.1;
        195.111.254.1;
    }
    radius-server {
        195.111.253.2 {
            secret "<encrypted secret key>"; # SECRET-DATA
            timeout 1;
            rctry 2;
        }
        195.111.254.2 {
            secret "<encrypted secret key>"; # SECRET-DATA
            timeout 1;
            retry 2;
        }
    }
    login {
        message "\n         +---------------------------------------+ \n          |
 |\n          | ********* W A R N I N G ********* |\n          |
 |\n          |   STRICTLY NO UNAUTHORISED ACCESS   |\n          |
 |\n          +---------------------------------------+\n\n";
        class operator {
            permissions [ admin configure interface interface-control network
                routing snmp system trace view firewall ];
            allow-commands "clear interfaces statistics";
        }
        class mysuperuser {
            idle-timeout 60;
            permissions all;
        }
```

```
        class viewall {
            idle-timeout 60;
            permissions [ admin interface network routing snmp system trace
                view firewall ];
        }
        user mysuperuser {
            full-name "Template RADIUS Account";
            uid 1000;
            class mysuperuser;
        }
        user operator {
            full-name "Template RADIUS Account";
            uid 1001;
            class operator;
        }
        user viewall {
            full-name "Template RADIUS Account";
            uid 1002;
            class viewall;
        }
        user admin {
            full-name "Last resort account";
            uid 9999;
            class mysuperuser;
            authentication {
                encrypted-password "<encrypted password>";
            }
        }
    }
    services {
        inactive: ftp {
            connection-limit 1;
            rate-limit 1;
        }
        ssh {
            root-login deny;
            protocol-version 2;
            connection-limit 4;
            rate-limit 4;
        }
        inactive: telnet {
            connection-limit 4;
            rate-limit 4;
        }
    }
    syslog {
        archive size 10m files 10;
        user * {
            any emergency;
        }
        host 195.111.253.3 {
            any info;
        }
        host 195.111.254.3 {
```

```
                any info;
            }
            file messages {
                any info;
                authorization info;
            }
            file security.log {
                interactive-commands any;
                archive size 10m files 10;
            }
        }
    }
    ntp {
        boot-server 195.111.253.4;
        server 195.111.253.4;
        server 195.111.254.4;
    }
}
chassis {
    no-reset-on-timeout;
    dump-on-panic;
    redundancy {
        routing-engine 0 master;
        routing-engine 1 backup;
        failover on-loss-of-keepalives;
        keepalive-time 300;
    }
    fpc 0 {
        pic 0 {
            framing sdh;
        }
    }
}
interfaces {
    apply-groups sdh-template;
    so-0/0/0 {
        description "Direct Peering Link to Peer1";
        clocking external;
        encapsulation cisco-hdlc;
        unit 0 {
            description "Link to Peer1 Lon-Br1 POS1/0";
            family inet {
                address 195.111.199.1/30;
            }
        }
    }
    ge-0/1/0 {
        description "GigE to primary local LAN";
        unit 0 {
            family inet {
                address 195.111.100.201/24;
            }
            family iso;
            family mpls;
        }
```

```
        }
    ge-1/1/0 {
        description "GigE to backup local LAN";
        unit 0 {
            family inet {
                address 195.111.101.201/24;
            }
            family iso;
            family mpls;
        }
    }
    ge-1/3/0 {
        description "GigE to Internet Exchange LAN";
        unit 0 {
            family inet {
                address 158.175.0.100/24;
            }
        }
    }
    lo0 {
        unit 0 {
            family inet {
                filter {
                    input accessfilter;
                }
                address 195.111.255.1/32 {
                    primary;
                    preferred;
                }
                address 195.111.255.254/32;
            }
            family iso {
                address 49.0000.1921.6825.5001.00;
            }
        }
    }
}
forwarding-options {
    sampling {
        input {
            family inet {
                rate 500;
            }
        }
        output {
            cflowd 195.111.254.5 {
                version 5;
                local-dump;
                autonomous-system-type origin;
            }
        }
    }
}
snmp {
```

```
        description "London-Border1 Juniper M10i";
        location "London";
        contact "Netops +44 1999 999999";
        community <string> {
            authorization read-only;
            clients {
                195.111.253.5/32;
                195.111.253.6/32;
            }
        }
        trap-options {
            source-address lo0;
        }
        trap-group Juniper {
            version v2;
            categories {
                authentication;
                chassis;
                link;
                routing;
                startup;
            }
            targets {
                195.111.253.5;
                195.111.253.6;
            }
        }
    }
routing-options {
    static {
        route 10.0.0.0/8 {
            next-hop 10.0.0.254;
            retain;
            no-readvertise;
        }
    }
    autonomous-system 1;
    forwarding-table {
        export load-share;
        unicast-reverse-path feasible-paths;
    }
    interface-routes {
        rib-group inet interface-rib;
    }
    rib-groups {
        bgp-rib {
            import-rib [ inet.0 inet.2 ];
            export-rib [ inet.0 ];
        }
        multicast-rib {
            import-rib [ inet.2 ];
            export-rib [ inet.2 ];
        }
        interface-rib {
```

```
                import-rib [ inet.0 inet.2 ];
            }
        }
}
protocols {
    mpls {
        traceoptions {
            file mpls.log files 10 size 1m;
            flag hello detail;
            flag state detail;
        }
        log-updown;
        no-propagate-ttl;
        interface ge-0/1/0.0;
        interface ge-1/1/0.0;
    }
    ldp {
        traceoptions {
            file ldp.log files 10 size 1m;
            flag hello detail;
            flag state detail;
        }
        log-updown;
        interface ge-0/1/0.0;
        interface ge-1/1/0.0;
        interface lo0.0;
        interface fxp0.0 {
            disable;
        }
    }
    bgp {
        traceoptions {
            file bgp.log files 10 size 1m;
            flag open detail;
            flag state detail;
        }
        rib-group bgp-rib;
        log-updown;
        group LONDON-RR-CLUSTER {
            type internal;
            export next-hop-self;
            local-address 195.111.255.201;
            neighbor 195.111.255.1;
            neighbor 195.111.255.2;
        }
        group PEER {
            type external;
            import [ 28bitfilter peer-in ];
            export transit-out;
            neighbor 195.111.126.2;
        }
    }
    isis {
        traceoptions {
```

```
                file isis.log files 10 size 1m;
                flag hello detail;
                flag error detail;
                flag state detail;
            }
        multicast-topology;
        interface ge-0/1/0.0 {
            level 1 {
                metric 25;
                hello-interval 10;
                hold-time 40;
            }
            level 2 disable;
        }
        interface ge-1/1/0.0 {
            level 1 {
                metric 30;
                hello-interval 10;
                hold-time 40;
            }
            level 2 disable;
        }
        interface lo0.0;
        interface fxp0.0 {
            disable;
        }
    }
    pim {
        traceoptions {
            file pim.log files 10 size 1m;
            flag hello detail;
            flag join detail;
            flag rp detail;
            flag assert detail;
            flag state detail;
        }
        rib-group multicast-rib;
        rp {
            static {
                address 195.111.255.254 {
                    group-ranges {
                        224.0.0.0/4;
                    }
                }
            }
        }
        interface so-0/0/0.0 {
            mode sparse;
        }
        interface ge-0/1/0.0{
            mode sparse;
        }
        interface ge-1/1/0.0{
            mode sparse;
```

```
        }
        interface ge-1/3/0.0{
            mode sparse;
        }
        interface lo0.0{
            mode sparse;
        }
        interface fxp0.0 {
            disable;
        }
    }
}
policy-options {
    prefix-list my-martians {
        0.0.0.0/7;
        2.0.0.0/8;
        5.0.0.0/8;
        7.0.0.0/8;
        10.0.0.0/8;
        23.0.0.0/8;
        27.0.0.0/8;
        31.0.0.0/8;
        36.0.0.0/7;
        39.0.0.0/8;
        41.0.0.0/8;
        42.0.0.0/8;
        49.0.0.0/8;
        50.0.0.0/8;
        58.0.0.0/7;
        70.0.0.0/7;
        72.0.0.0/5;
        83.0.0.0/8;
        84.0.0.0/6;
        88.0.0.0/5;
        96.0.0.0/3;
        169.254.0.0/16;
        172.16.0.0/12;
        173.0.0.0/8;
        174.0.0.0/7;
        176.0.0.0/5;
        184.0.0.0/6;
        189.0.0.0/8;
        190.0.0.0/8;
        192.0.2.0/24;
        192.168.0.0/16;
        197.0.0.0/8;
        198.18.0.0/15;
        223.0.0.0/8;
        224.0.0.0/3;
    }
    prefix-list our-aggregates {
        195.111.0.0/16;
    }
    policy-statement load-share {
```

```
        term load-balance-per-flow {
            then {
                load-balance per-packet;
            }
        }
    }
    policy-statement 28-bit-filter {
        term block-0-7 {
            from {
                protocol bgp;
                route-filter 0.0.0.0/0 upto /7;
            }
            then reject;
        }
        term permit-8-to-28 {
            from {
                protocol bgp;
                route-filter 0.0.0.0/0 upto /28;
            }
            then accept;
        }
        term block-29-32 {
            from {
                protocol bgp;
                route-filter 0.0.0.0/0 upto /32;
            }
            then reject;
        }
    }
    policy-statement next-hop-self {
        term set-next-hop-self {
            from protocol bgp;
            then next-hop-self;
        }
    }
    policy-statement AS5555-in {
        term drop-illegal-prefixes {
            from {
                protocol bgp;
                prefix-list my-martians;
            }
            then reject;
        }
        term accept-and-mark {
            from {
                protocol bgp;
                as-path AS5555-list;
            }
            then {
                community add Peer;
                accept;
            }
        }
    }
}
```

```
    policy-statement transit-out {
        term block-peer-routes {
            from {
                protocol bgp;
                community Peer;
            }
            then reject;
        }
        term block-transit-routes {
            from {
                protocol bgp;
                community Transit;
            }
            then reject;
        }
        term send-our-aggregates {
            from {
                protocol bgp;
                community OurAggregates;
            }
            then accept;
        }
        term leak-dual-homed-customer-routes {
            from {
                protocol bgp;
                community PleaseLeakThis;
            }
            then accept;
        }
        term block-more-specifics {
            from {
                protocol bgp;
                prefix-list our-aggregates;
            }
            then reject;
        }
        term send-customer-routes {
            from {
                protocol bgp;
                community Customer;
            }
            then accept;
        }
    }
    as-path AS5555-list "5555+ (6666|7777)?";
    community Customer members 1:1;
    community Peer members 1:2;
    community Transit members 1:3;
    community DontLeakThis members 1:999;
    community PleaseLeakThis members 1:1000;
    community OurAggregates members 1:50000;
}
class-of-service {
    classifiers {
```

```
        exp mpls-exp {
            forwarding-class expedited-forwarding {
                loss-priority high code-points [ 101 110 111 ];
            }
            forwarding-class assured-forwarding {
                loss-priority high code-points 011;
            }
            forwarding-class best-effort {
                loss-priority high code-points [ 000 010 100 ];
            }
            forwarding-class less-than-best-effort {
                loss-priority high code-points 001;
            }
        }
        dscp diffserv {
            forwarding-class expedited-forwarding {
                loss-priority high code-points [ ef cs6 cs7 cs5 ];
            }
            forwarding-class assured-forwarding {
                loss-priority high code-points af31;
                loss-priority low code-points [ af32 af33 ];
            }
            forwarding-class best-effort {
                loss-priority high code-points [ be cs2 cs4 ];
            }
            forwarding-class less-than-best-effort {
                loss-priority high code-points cs1;
            }
        }
        inet-precedence inet-prec {
            forwarding-class expedited-forwarding {
                loss-priority high code-points [ 101 110 111 ];
            }
            forwarding-class assured-forwarding {
                loss-priority high code-points 011;
            }
            forwarding-class best-effort {
                loss-priority high code-points [ 000 010 100 ];
            }
            forwarding-class less-than-best-effort {
                loss-priority high code-points 001;
            }
        }
    }
    drop-profiles {
        aggressive {
            fill-level 10 drop-probability 30;
            fill-level 70 drop-probability 100;
        }
        gentle {
            fill-level 20 drop-probability 20;
            fill-level 80 drop-probability 80;
        }
    }
```

```
forwarding-classes {
    queue 0 best-effort;
    queue 1 less-than-best-effort;
    queue 2 assured-forwarding;
    queue 3 expedited-forwarding;
}
interfaces {
    so-0/*/* {
        scheduler-map diffserv;
        unit * {
            classifiers {
                dscp diffserv;
                exp mpls-exp;
            }
        }
    }
    ge-*/1/0 {
        scheduler-map diffserv;
        unit * {
            classifiers {
                inet-precedence nondiffserv;
                exp mpls-exp;
            }
        }
    }
    ge-*/3/0 {
        scheduler-map diffserv;
        unit * {
            classifiers {
                inet-precedence inet-prec;
            }
        }
    }
}
scheduler-maps {
    diffserv {
        forwarding-class expedited-forwarding scheduler expedited-
            forwarding;
        forwarding-class assured-forwarding scheduler assured-forwarding;
        forwarding-class best-effort scheduler best-effort;
        forwarding-class less-than-best-effort scheduler less-than-
            best-effort;
    }
}
schedulers {
    expedited-forwarding {
        transmit-rate percent 10 exact;
        buffer-size percent 10;
        priority strict-high;
    }
    assured-forwarding {
        transmit-rate percent 10;
        buffer-size percent 10;
        priority high;
```

```
                drop-profile-map loss-priority high protocol any drop-profile
                    aggressive;
                drop-profile-map loss-priority low protocol any drop-profile
                    gentle;
            }
        best-effort {
            transmit-rate percent 79;
            buffer-size percent 79;
            priority low;
            drop-profile-map loss-priority any protocol any drop-profile
                gentle;
        }
        less-than-best-effort {
            transmit-rate remainder;
            buffer-size percent 1;
            priority low;
            drop-profile-map loss-priority any protocol any drop-profile
                gentle;
        }
    }
}
}
firewall {
    filter accessfilter {
        term allow-telnet-from-very-small-list {
            from {
                source-address {
                    195.111.200.0/24;
                    195.111.255.0/24;
                }
                protocol tcp;
                destination-port telnet;
            }
            then accept;
        }
        term deny-telnet-from-everywhere-else {
            from {
                protocol tcp;
                destination-port telnet;
            }
            then {
                log;
                reject;
            }
        }
        term allow-ftp-from-very-small-list {
            from {
                source-address {
                    195.111.200.0/24;
                }
                protocol tcp;
                port [ ftp ftp-data ];
            }
            then accept;
        }
```

```
            term deny-ftp-from-everywhere-else {
                from {
                    protocol tcp;
                    destination-port [ ftp ftp-data ];
                }
                then {
                    log;
                    reject;
                }
            }
            term allow-everything-else {
                then accept;
            }
        }
        filter cflowd {
            term sample-and-accept {
                then {
                    sample;
                    accept;
                }
            }
        }
    }
}
```

12.6   BORDER ROUTER RUNNING MPLS WITH LDP, BGP POLICY TO PEERS, MBGP, PIM-SM (IOS)

```
no service tcp-small-servers
no service udp-small-servers
service password-encryption
no service pad
no ip finger
no ip bootp server
!
aaa new-model
aaa authentication login use-radius group radius local
aaa authorization exec default group radius
aaa accounting commands 0 default start-stop group radius
aaa accounting system default start-stop group radius
aaa accounting connection default start-stop group radius
!
enable secret 5 XXXXXXXXX
!
username admin password 7 XXXXXXXXX
hostname London-Border1
ip domain-name exampledomain.com
ip name-server 195.111.253.1 195.111.254.1
!
```

```
ip ftp username robot-user
ip ftp password 0 robot-pwd
exception protocol ftp
exception dump 195.111.253.8
exception core-file London-Border1-corefile
!
ip cef distributed
!
! Enable MPLS
mpls ip
mpls ldp advertise-labels for 1
mpls ldp router-id lo0
mpls ldp discovery targeted-hellos accept
!
ip multicast-routing
!
logging 195.111.253.3 6
logging 195.111.254.3 6
logging console 0
logging source-interface lo0
!
ip flow-export 195.111.254.5 9995 version 5 origin-as
!
interface POS0/0
 description Direct link to Peer 1
 encapsulation hdlc
 ip address 195.111.199.1 255.255.255.252
 pos framing sdh
 pos flag s1s0 2
 pos scramble-atm
 crc 32
 clock source line
!
interface GigabitEthernet1/0
 description GigE peering link
 ip address 158.175.0.100 255.255.255.0
!
interface GigabitEthernet2/0
 description GigE to primary local LAN
 ip address 195.111.100.201 255.255.255.0
 ip access-group accessfilter in
 service-policy output core-forwarding
 ip router isis
 isis circuit-type level-1
 isis metric 25 level-1
 isis hello-interval 10 level-1
 isis hello-multiplier 4
 mpls ip
 mpls label protocol ldp
 ip pim sparse-mode
!
interface GigabitEthernet3/0
 description GigE to secondary local LAN
 ip address 195.111.101.201 255.255.255.0
```

```
 ip access-group accessfilter in
 service-policy output core-forwarding
 ip router isis
 isis circuit-type level-1
 isis metric 30 level-1
 isis hello-interval 10 level-1
 isis hello-multiplier 4
 mpls ip
 mpls label protocol ldp
 ip pim sparse-mode
!
interface Ethernet 4/0/0
 description Link to Management Network
 ip address 10.0.0.1 255.255.255.0
 ip access-group mgmt-in in
 ip access-group mgmt-out out
!
interface POS6/0
 description POS link to SanFrancisco-Core1 POS6/0
 ip address 195.111.1.9 255.255.255.252
 ip access-group accessfilter in
 pos framing sdh
 pos flag s1s0 2
 pos scramble-atm
 crc 32
 clock source line
 service-policy output core-forwarding
 ip router isis
 isis circuit-type level-2-only
 isis metric 55 level-2
 mpls ip
 mpls traffic-eng tunnels
 ip pim sparse-mode
!
interface POS7/0
 description POS link to Boston-Core1 POS7/0
 ip address 195.111.1.13 255.255.255.252
 ip access-group accessfilter in
 pos framing sdh
 pos flag s1s0 2
 pos scramble-atm
 crc 32
 clock source line
 service-policy output core-forwarding
 ip router isis
 isis circuit-type level-2-only
 isis metric 55 level-2
 mpls ip
 mpls traffic-eng tunnels
 ip pim sparse-mode
!
interface lo0
 description Loopback
 ip address 195.111.255.1 255.255.255.255
```

```
 ip router isis
!
ip access-list 1 permit 195.111.255.0 0.0.0.255
ip access-list 99 permit host 195.111.253.5
ip access-list 99 permit host 195.111.253.6
ip access-list extended mgmt-in
 permit 10.0.0.0 0.0.0.255 host 195.111.255.1
 permit 10.0.0.0 0.0.0.255 host 10.0.0.1
 deny any any
ip access-list extended mgmt-out
 permit host 195.111.255.1 10.0.0.0 0.255.255.255
 permit host 10.0.0.1 10.0.0.0 0.255.255.255
 deny any any
ip access-list extended accessfilter
 permit ip 195.111.250.0 0.0.0.255 host 195.111.255.1 eq ssh
 permit ip 195.111.250.0 0.0.0.255 195.111.0.0 0.0.3.255 eq ssh
 permit ip 195.111.255.0 0.0.0.255 host 195.111.255.1 eq ssh
 permit ip 195.111.255.0 0.0.0.255 195.111.0.0 0.0.3.255 eq ssh
 permit ip 10.24.0.0 0.0.0.255 host 195.111.255.1 eq ssh
 permit ip 10.24.0.0 0.0.0.255 195.111.0.0 0.0.3.255 eq ssh
 deny ip any host 195.111.255.1 eq ssh
 deny ip any 195.111.0.0 0.0.3.255 eq ssh
 permit ip 195.111.250.0 0.0.0.255 host 195.111.255.1 eq ftp
 permit ip 195.111.250.0 0.0.0.255 host 195.111.255.1 eq ftp-data
 permit ip 195.111.255.0 0.0.0.255 host 195.111.255.1 eq ftp
 permit ip 195.111.255.0 0.0.0.255 host 195.111.255.1 eq ftp-data
 permit ip 195.111.250.0 0.0.0.255 195.111.0.0 0.0.3.255 eq ftp
 permit ip 195.111.250.0 0.0.0.255 195.111.0.0 0.0.3.255 eq ftp-data
 permit ip 195.111.255.0 0.0.0.255 195.111.0.0 0.0.3.255 eq ftp
 permit ip 195.111.255.0 0.0.0.255 195.111.0.0 0.0.3.255 eq ftp-data
 deny ip any host 195.111.255.1 eq ftp
 deny ip any host 195.111.255.1 eq ftp data
 deny ip any 195.111.0.0 0.0.3.255 eq ftp
 deny ip any 195.111.0.0 0.0.3.255 eq ftp
 permit ip 195.111.250.0 0.0.0.255 host 195.111.255.1 eq telnet
 permit ip 195.111.250.0 0.0.0.255 195.111.0.0 0.0.3.255 eq telnet
 permit ip 195.111.255.0 0.0.0.255 host 195.111.255.1 eq telnet
 permit ip 195.111.255.0 0.0.0.255 195.111.0.0 0.0.3.255 eq telnet
 deny ip any host 195.111.255.1 eq ssh
 deny ip any 195.111.0.0 0.0.3.255 eq ssh
 permit ip any any
!
ip as-path access-list 1 permit ^5555(_5555)*$
ip as-path access-list 1 permit ^5555(_5555)*_6666$
ip as-path access-list 1 permit ^5555(_5555)*_7777$
!
ip prefix-list my-martians deny 0.0.0.0/7
ip prefix-list my-martians deny 2.0.0.0/8
ip prefix-list my-martians deny 5.0.0.0/8
ip prefix-list my-martians deny 7.0.0.0/8
ip prefix-list my-martians deny 10.0.0.0/8
ip prefix-list my-martians deny 23.0.0.0/8
ip prefix-list my-martians deny 27.0.0.0/8
ip prefix-list my-martians deny 31.0.0.0/8
```

```
ip prefix-list my-martians deny 36.0.0.0/7
ip prefix-list my-martians deny 39.0.0.0/8
ip prefix-list my-martians deny 41.0.0.0/8
ip prefix-list my-martians deny 42.0.0.0/8
ip prefix-list my-martians deny 49.0.0.0/8
ip prefix-list my-martians deny 50.0.0.0/8
ip prefix-list my-martians deny 58.0.0.0/7
ip prefix-list my-martians deny 70.0.0.0/7
ip prefix-list my-martians deny 72.0.0.0/5
ip prefix-list my-martians deny 83.0.0.0/8
ip prefix-list my-martians deny 84.0.0.0/6
ip prefix-list my-martians deny 88.0.0.0/5
ip prefix-list my-martians deny 96.0.0.0/3
ip prefix-list my-martians deny 169.254.0.0/16
ip prefix-list my-martians deny 172.16.0.0/12
ip prefix-list my-martians deny 173.0.0.0/8
ip prefix-list my-martians deny 174.0.0.0/7
ip prefix-list my-martians deny 176.0.0.0/5
ip prefix-list my-martians deny 184.0.0.0/6
ip prefix-list my-martians deny 189.0.0.0/8
ip prefix-list my-martians deny 190.0.0.0/8
ip prefix-list my-martians deny 192.0.2.0/24
ip prefix-list my-martians deny 192.168.0.0/16
ip prefix-list my-martians deny 197.0.0.0/8
ip prefix-list my-martians deny 198.18.0.0/15
ip prefix-list my-martians deny 223.0.0.0/8
ip prefix-list my-martians deny 224.0.0.0/3
ip prefix-list my-martians permit 0.0.0.0/0 ge 8 le 28
!
ip community-list standard Customer permit 1:1
ip community-list standard Peer permit 1:2
ip community-list standard Transit permit 1:3
ip community-list standard DontLeakThis permit 1:999
ip community-list standard PleaseLeakThis permit 1:1000
!
class-map match-any expedited-forwarding
 match ip precedence 101 110 111
 match mpls experimental 101 110 111
 match ip dscp ef cs5 cs6 cs7
class-map match-any assured-forwarding
 match ip precedence 011
 match mpls experimental 011
 match ip dscp af31 af32 af33
class-map match-any best-effort
 match ip precedence 000 010 100
 match mpls experimental 000 010 100
 match ip dscp default cs2 cs4
class-map match-any less-than-best-effort
 match ip precedence 001
 match mpls experimental 001
 match ip dscp cs1
!
policy-map core-forwarding
 class expedited-forwarding
```

```
 bandwidth percent 10
 class assured-forwarding
 bandwidth percent 10
 class best-effort
 bandwidth percent 79
 class less-than-best-effort
 bandwidth percent 1
!
router isis
 net 49.0000.1921.6825.5201.00
 is-type level-2
!
router bgp 1
 no synchronize
 neighbor LONDON-RR-CLUSTER peer-group
 neighbor LONDON-RR-CLUSTER remote-as 1
 neighbor LONDON-RR-CLUSTER send-community
 neighbor LONDON-RR-CLUSTER update-source loopback0
 neighbor LONDON-RR-CLUSTER next-hop-self
 neighbor PEER peer-group
 neighbor PEER send-community
 neighbor PEER prefix-list my-martians in
 neighbor PEER prefix-list my-martians out
 neighbor PEER route-map transit-out out
 neighbor 195.111.255.1 peer-group LONDON-RR-CLUSTER
 neighbor 195.111.255.2 peer-group LONDON-RR-CLUSTER
 neighbor 195.111.125.2 peer-group PEER
 neighbor 195.111.125.2 remote-as 5555
 neighbor 195.111.125.2 route-map AS5555-in in
 no auto-summary
!
ip bgp-community new-format
!
route-map AS5555-in permit 10
 match as path 1
 set community Peer additive
!
route-map transit-out permit 10
 match community PleaseLeakThis
!
route-map transit-out permit 20
 match community Customer
!
route-map transit-out permit 30
 match community OurAggregate
!
ip pim rp-address 195.111.255.254
!
aaa group server my-radius-group
  server host 195.111.253.2 auth-port 1812 acct-port 1813 key <string>
  server host 195.111.254.2 auth-port 1812 acct-port 1813 key <string>
!
snmp-server community <string> ro 99
snmp-server host 195.111.253.5 traps version 2c <string>
```

```
snmp-server host 195.111.253.6 traps version 2c <string>
snmp-server enable traps
snmp-server trap-source loopback0
snmp-server contact Netops +44 1999 999999
snmp-server location London
!
ntp bootserver 195.111.253.4
ntp server 195.111.253.4
ntp server 195.111.254.4
!
no ip http server
no ip http server-secure
!
login banner ^C
            +-------------------------------------+
            |                                     |
            | ********** W A R N I N G ********** |
            |                                     |
            |   STRICTLY NO UNAUTHORISED ACCESS   |
            |                                     |
            +-------------------------------------+
^C
!
line con 0
 login authentication use-radius
 exec-timeout 0 0
 transport preferred none
line vty 0 4
 login authentication use-radius
 exec-timeout 0 0
 transport preferred none
```

12.7 TRANSIT ROUTER RUNNING MPLS WITH LDP, BGP POLICY TO UPSTREAM TRANSIT PROVIDERS, MBGP, PIM-SM (JUNOS)

```
groups {
    re0 {
        system {
            host-name London-Transit1-re0;
        }
        interface {
            fxp0 {
                unit 0 {
                    family inet {
                        address 10.0.0.210/24;
                    }
                }
            }
        }
```

```
        }
    re1 {
        system {
            host-name London-Transit1-re1;
        }
        interface {
            fxp0 {
                unit 0 {
                    family inet {
                        address 10.0.0.211/24;
                    }
                }
            }
        }
    }
    sdh-template {
        interfaces {
            <so-*/*/*> {
                sonet-options {
                    rfc-2615;
                }
            }
        }
    }
}
apply-groups [ re0 re1 ];
system {
    domain-name exampledomain.com;
    time-zone Europe/London;
    no-redirects;
    dump-on-panic;
    mirror-flash-on-disk;
    backup-router 10.0.0.254 destination 10.0.0.0/8;
    authentication-order radius;
    ports {
        console type vt100;
    }
    archival configuration {
        transfer-interval 1440;
        transfer-on-commit;
        archive-sites {
            ftp://robot-user:robot-pwd@195.111.253.8/data/configs;
        }
    }
    root-authentication {
        encrypted-password "<encrypted password>"; # SECRET-DATA
    }
    name-server {
        195.111.253.1;
        195.111.254.1;
    }
    radius-server {
        195.111.253.2 {
            secret "<encrypted secret key>"; # SECRET-DATA
```

```
            timeout 1;
            retry 2;
        }
        195.111.254.2 {
            secret "<encrypted secret key>"; # SECRET-DATA
            timeout 1;
            retry 2;
        }
    }
    login {
        message "\n        +-------------------------------------+ \n        |
|\n        | ********** W A R N I N G ********** |\n            |
|\n        |    STRICTLY NO UNAUTHORISED ACCESS   |\n            |
|\n        +-------------------------------------+\n\n";
        class operator {
            permissions [ admin configure interface interface-control network
                routing snmp system trace view firewall ];
            allow-commands "clear interfaces statistics";
        }
        class mysuperuser {
            idle-timeout 60;
            permissions all;
        }
        class viewall {
            idle-timeout 60;
            permissions [ admin interface network routing snmp system trace
                view firewall ];
        }
        user operator {
            full-name "Template RADIUS Account";
            uid 1001;
            class operator;
        }
        user mysuperuser {
            full-name "Template RADIUS Account";
            uid 1000;
            class mysuperuser;
        }
        user viewall {
            full-name "Template RADIUS Account";
            uid 1002;
            class viewall;
        }
        user admin {
            full-name "Last resort account";
            uid 9999;
            class mysuperuser;
            authentication {
                encrypted-password "<encrypted password>";
            }
        }
    }
    services {
        inactive: ftp {
```

```
                connection-limit 1;
                rate-limit 1;
            }
            ssh {
                root-login deny;
                protocol-version 2;
                connection-limit 4;
                rate-limit 4;
            }
            inactive: telnet {
                connection-limit 4;
                rate-limit 4;
            }
        }
    syslog {
        archive size 10m files 10;
        user * {
            any emergency;
        }
        host 195.111.253.3 {
            any info;
        }
        host 195.111.254.3 {
            any info;
        }
        file messages {
            any info;
            authorization info;
        }
        file security.log {
            interactive-commands any;
            archive size 10m files 10,
        }
    }
    ntp {
        boot-server 195.111.253.4;
        server 195.111.253.4;
        server 195.111.254.4;
    }
}
chassis {
    no-reset-on-timeout;
    dump-on-panic;
    redundancy {
        routing-engine 0 master;
        routing-engine 1 backup;
        failover on-loss-of-keepalives;
        keepalive-time 300;
    }
    fpc 0 {
        pic 0 {
            framing sdh;
        }
    }
```

```
    fpc 1 {
        pic 0 {
            framing sdh;
        }
    }
}
interfaces {
    apply-groups sdh-template;
    so-0/0/0 {
        description "Link to Transit Provider";
        clocking external;
        encapsulation ppp;
        unit 0 {
            description "Link to Transit1 Lon-AR1 so-0/0/0.0";
            family inet {
                address 195.111.125.1/30;
            }
        }
    }
    ge-0/1/0 {
        description "GigE to primary local LAN";
        unit 0 {
            family inet {
                address 195.111.100.210/24;
            }
            family iso;
            family mpls;
        }
    }
    so-1/0/0 {
        description "Link to Transit Provider";
        clocking external;
        encapsulation ppp;
        unit 0 {
            description "Link to Transit1 Lon-AR2 so-0/0/0.0";
            family inet {
                address 195.111.125.5/30;
            }
        }
    }
    ge-1/1/0 {
        description "GigE to backup local LAN";
        unit 0 {
            family inet {
                address 195.111.101.210/24;
            }
            family iso;
            family mpls;
        }
    }
    lo0 {
        unit 0 {
            family inet {
                /* filtering on the loopback protects all interfaces */
```

```
                    filter {
                        input accessfilter;
                    }
                    address 195.111.255.210/32 {
                        primary;
                        preferred;
                    }
                }
                family iso {
                    address 49.0000.1921.6825.5210.00;
                }
            }
        }
    }
}
forwarding-options {
    sampling {
        input {
            family inet {
                rate 500;
            }
        }
        output {
            cflowd 195.111.254.5 {
                version 5;
                local-dump;
                autonomous-system-type origin;
            }
        }
    }
}
snmp {
    description "London Transit1 Juniper M10i";
    location "London";
    contact "Netops +44 1999 999999";
    community <string> {
        authorization read-only;
        clients {
            195.111.253.5/32;
            195.111.253.6/32;
        }
    }
    trap-options {
        source-address lo0;
    }
    trap-group Juniper {
        version v2;
        categories {
            authentication;
            chassis;
            link;
            routing;
            startup;
        }
        targets {
```

```
                195.111.253.5;
                195.111.253.6;
            }
        }
    }
routing-options {
    static {
        route 10.0.0.0/8 {
            next-hop 10.0.0.254;
            retain;
            no-readvertise;
        }
    }
    autonomous-system 1;
    forwarding-table {
        export load-share;
    }
    interface-routes {
        rib-group inet interface-rib;
    }
    rib-groups {
        bgp-rib {
            import-rib [ inet.0 inet.2 ];
            export-rib [ inet.0 ];
        }
        multicast-rib {
            import-rib [ inet.2 ];
            export-rib [ inet.2 ];
        }
        interface-rib {
            import-rib [ inet.0 inet.2 ];
        }
    }
}
protocols {
    mpls {
        traceoptions {
            file mpls.log files 10 size 1m;
            flag hello detail;
            flag state detail;
        }
        apply-groups mpls-template;
        log-updown;
        no-propagate-ttl;
        interface ge-0/1/0.0;
        interface ge-1/1/0.0;
        interface lo0.0;
        interface fxp0.0 {
            disable;
        }
    }
    ldp {
        traceoptions {
            file ldp.log files 10 size 1m;
```

```
                flag hello detail;
                flag state detail;
            }
            log-updown;
            interface ge-0/1/0.0;
            interface ge-1/1/0.0;
            interface lo0.0;
            interface fxp0.0 {
                disable;
            }
        }
        bgp {
            traceoptions {
                file bgp.log files 10 size 1m;
                flag open detail;
                flag state detail;
            }
            rib-group bgp-rib;
            log-updown;
            group LONDON-RR-CLUSTER {
                type internal;
                local-address 195.111.255.210;
                neighbor 195.111.255.1;
                neighbor 195.111.255.2;
            }
            group TRANSIT-PROVIDER {
                type external;
                import [ 28-bit-filter transit-in ];
                export [ transit-out ];
                neighbor 195.111.125.2;
                neighbor 195.111.125.6;
                peer-as 9999;
            }
        }
        isis {
            traceoptions {
                file isis.log files 10 size 1m;
                flag hello detail;
                flag error detail;
                flag state detail;
            }
            multicast-topology;
            interface ge-0/1/0.0 {
                level 1 {
                    metric 25;
                    hello-interval 10;
                    hold-time 40;
                }
                level 2 disable;
            }
            interface ge-1/1/0.0 {
                level 1 {
                    metric 30;
                    hello-interval 10;
```

```
                hold-time 40;
            }
            level 2 disable;
        }
        interface lo0.0;
        interface fxp0.0 {
            disable;
        }
    }
    pim {
        traceoptions {
            file pim.log files 10 size 1m;
            flag hello detail;
            flag join detail;
            flag rp detail;
            flag assert detail;
            flag state detail;
        }
        rib-group multicast-rib;
        rp {
            static {
                address 195.111.255.254 {
                    group-ranges {
                        224.0.0.0/4;
                    }
                }
            }
        }
        interface so-0/0/0.0 {
            mode sparse;
        }
        interface ge-2/0/0.0{
            mode sparse;
        }
        interface ge-3/0/0.0{
            mode sparse;
        }
        interface lo0.0{
            mode sparse;
        }
        interface fxp0.0 {
            disable;
        }
    }
}
policy-options {
    prefix-list my-martians {
        0.0.0.0/7;
        2.0.0.0/8;
        5.0.0.0/8;
        7.0.0.0/8;
        10.0.0.0/8;
        23.0.0.0/8;
        27.0.0.0/8;
```

```
            31.0.0.0/8;
            36.0.0.0/7;
            39.0.0.0/8;
            41.0.0.0/8;
            42.0.0.0/8;
            49.0.0.0/8;
            50.0.0.0/8;
            58.0.0.0/7;
            70.0.0.0/7;
            72.0.0.0/5;
            83.0.0.0/8;
            84.0.0.0/6;
            88.0.0.0/5;
            96.0.0.0/3;
            169.254.0.0/16;
            172.16.0.0/12;
            173.0.0.0/8;
            174.0.0.0/7;
            176.0.0.0/5;
            184.0.0.0/6;
            189.0.0.0/8;
            190.0.0.0/8;
            192.0.2.0/24;
            192.168.0.0/16;
            197.0.0.0/8;
            198.18.0.0/15;
            223.0.0.0/8;
            224.0.0.0/3;
        }
        prefix-list our-aggregates {
            195.111.0.0/16;
        }
        policy-statement 28-bit-filter {
            term block-0-7 {
                from {
                    protocol bgp;
                    route-filter 0.0.0.0/0 upto /7;
                }
                then reject;
            }
            term permit-8-to-28 {
                from {
                    protocol bgp;
                    route-filter 0.0.0.0/0 upto /28;
                }
                then accept;
            }
            term block-29-32 {
                from {
                    protocol bgp;
                    route-filter 0.0.0.0/0 upto /32;
                }
                then reject;
            }
```

```
            }
    policy-statement load-share {
        term load-balance-per-flow {
            then {
                load-balance per-packet;
            }
        }
    }
    policy-statement next-hop-self {
        term set-next-hop-self {
            from protocol bgp;
            then next-hop-self;
        }
    }
    policy-statement transit-in {
        term drop-illegal-prefixes {
            from {
                protocol bgp;
                prefix-list my-martians;
            }
            then reject;
        }
        term accept-and-mark {
            from protocol bgp;
            then {
                community set Transit;
                accept;
            }
        }
    }
    policy-statement transit-out {
        term block-peer-routes {
            from {
                protocol bgp;
                community Peer;
            }
            then reject;
        }
        term block-transit-routes {
            from {
                protocol bgp;
                community Transit;
            }
            then reject;
        }
        term send-our-aggregates {
            from {
                protocol bgp;
                community OurAggregates;
            }
            then accept;
        }
        term leak-dual-homed-customer-routes {
            from {
```

```
                    protocol bgp;
                    community PleaseLeakThis;
                }
                then accept;
            }
        term block-more-specifics {
            from {
                protocol bgp;
                prefix-list our-aggregates;
            }
            then reject;
        }
        term send-customer-routes {
            from {
                protocol bgp;
                community Customer;
            }
            then accept;
        }
    }
    community Customer members 1:1;
    community Peer members 1:2;
    community Transit members 1:3;
    community DontLeakThis members 1:999;
    community PleaseLeakThis members 1:1000;
    community OurAggregates members 1:50000;
}
class-of-service {
    classifiers {
        exp mpls-exp {
            forwarding-class expedited-forwarding {
                loss-priority high code-points [ 101 110 111 ];
            }
            forwarding-class assured-forwarding {
                loss-priority high code-points 011;
            }
            forwarding-class best-effort {
                loss-priority high code-points [ 000 010 100 ];
            }
            forwarding-class less-than-best-effort {
                loss-priority high code-points 001;
            }
        }
        dscp diffserv {
            forwarding-class expedited-forwarding {
                loss-priority high code-points [ ef cs6 cs7 cs5 ];
            }
            forwarding-class assured-forwarding {
                loss-priority high code-points af31;
                loss-priority low code-points [ af32 af33 ];
            }
            forwarding-class best-effort {
                loss-priority high code-points [ be cs2 cs4 ];
            }
```

```
                forwarding-class less-than-best-effort {
                    loss-priority high code-points cs1;
                }
            }
            inet-precedence inet-prec {
                forwarding-class expedited-forwarding {
                    loss-priority high code-points [ 101 110 111 ];
                }
                forwarding-class assured-forwarding {
                    loss-priority high code-points 011;
                }
                forwarding-class best-effort {
                    loss-priority high code-points [ 000 010 100 ];
                }
                forwarding-class less-than-best-effort {
                    loss-priority high code-points 001;
                }
            }
        }
        drop-profiles {
            aggressive {
                fill-level 10 drop-probability 30;
                fill-level 70 drop-probability 100;
            }
            gentle {
                fill-level 20 drop-probability 20;
                fill-level 80 drop-probability 80;
            }
        }
        forwarding-classes {
            queue 0 best-effort;
            queue 1 less-than-best-effort;
            queue 2 assured-forwarding;
            queue 3 expedited-forwarding;
        }
        interfaces {
            so-0/*/* {
                scheduler-map diffserv;
                unit * {
                    classifiers {
                        dscp diffserv;
                        exp mpls-exp;
                    }
                }
            }
            ge-2/*/* {
                scheduler-map diffserv;
                unit * {
                    classifiers {
                        inet-precedence nondiffserv;
                        exp mpls-exp;
                    }
                }
            }
```

```
        ge-3/*/* {
            scheduler-map diffserv;
            unit * {
                classifiers {
                    inet-precedence nondiffserv;
                    exp mpls-exp;
                }
            }
        }
    }
    scheduler-maps {
        diffserv {
            forwarding-class expedited-forwarding scheduler
                expedited-forwarding;
            forwarding-class assured-forwarding scheduler assured-forwarding;
            forwarding-class best-effort scheduler best-effort;
            forwarding-class less-than-best-effort scheduler
                less-than-best-effort;
        }
    }
    schedulers {
        expedited-forwarding {
            transmit-rate percent 10 exact;
            buffer-size percent 10;
            priority strict-high;
        }
        assured-forwarding {
            transmit-rate percent 10;
            buffer-size percent 10;
            priority high;
            drop-profile-map loss priority high protocol any drop-profile
                aggressive;
            drop-profile-map loss-priority low protocol any drop-profile
                gentle;
        }
        best-effort {
            transmit-rate percent 79;
            buffer-size percent 79;
            priority low;
            drop-profile-map loss-priority any protocol any drop-profile
                gentle;
        }
        less-than-best-effort {
            transmit-rate remainder;
            buffer-size percent 1;
            priority low;
            drop-profile-map loss-priority any protocol any drop-profile
                gentle;
        }
    }
}
firewall {
    filter accessfilter {
```

```
        term allow-telnet-from-very-small-list {
            from {
                source-address {
                    195.111.200.0/24;
                    195.111.255.0/24;
                }
                protocol tcp;
                destination-port telnet;
            }
            then accept;
        }
        term deny-telnet-from-everywhere-else {
            from {
                protocol tcp;
                destination-port telnet;
            }
            then {
                log;
                reject;
            }
        }
        term allow-ftp-from-very-small-list {
            from {
                source-address {
                    195.111.200.0/24;
                }
                protocol tcp;
                port [ ftp ftp-data ];
            }
            then accept;
        }
        term deny-ftp-from-everywhere-else {
            from {
                protocol tcp;
                destination-port [ ftp ftp-data ];
            }
            then {
                log;
                reject;
            }
        }
        term allow-everything-else {
            then accept;
        }
    }
    filter cflowd {
        term sample-and-accept {
            then {
                sample;
                accept;
            }
        }
    }
}
```

12.8   TRANSIT ROUTER RUNNING MPLS WITH LDP, BGP POLICY TO UPSTREAM TRANSIT PROVIDERS, MBGP, PIM-SM (IOS)

```
no service tcp-small-servers
no service udp-small-servers
service password-encryption
no service pad
no ip finger
no ip bootp server
!
aaa new-model
aaa authentication login use-radius group radius local
aaa authorization exec default group radius
aaa accounting commands 0 default start-stop group radius
aaa accounting system default start-stop group radius
aaa accounting connection default start-stop group radius
!
enable secret 5 XXXXXXXXX
!
username admin password 7 XXXXXXXXX
hostname London-Transit1
ip domain-name exampledomain.com
ip name-server 195.111.253.1 195.111.254.1
!
ip ftp username robot-user
ip ftp password 0 robot-pwd
exception protocol ftp
exception dump 195.111.253.8
exception core-file London-Transit1-corefile
!
ip cef distributed
!
mpls ip
mpls ldp advertise-labels for 1
mpls ldp router-id lo0
mpls ldp discovery targeted-hellos accept
!
ip multicast-routing
!
logging 195.111.253.3 6
logging 195.111.254.3 6
logging console 0
logging source-interface lo0
!
ip flow-export 195.111.254.5 9995 version 5 origin-as
!
interface POS 0/0/0
 description Link to Transit Provider 1
 encapsulation ppp
```

```
 ip address 195.111.125.1 255.255.255.252
 ip access-group accessfilter in
 pos framing sdh
 pos flag s1s0 2
 pos scramble-atm
 crc 32
 clock source line
!
interface POS 1/0/0
 description Link to Transit Provider 2
 encapsulation ppp
 ip address 195.111.125.5 255.255.255.252
 access-group accessfilter in
 pos framing sdh
 pos flag s1s0 2
 pos scramble-atm
 crc 32
 clock source line
!
interface GigabitEthernet2/0/0
 description GigE to primary local LAN
 ip address 195.111.100.210 255.255.255.0
 ip access-group accessfilter in
 service-policy output core-forwarding
 ip router isis
 isis circuit-type level-1
 isis metric 25 level-1
 isis hello-interval 10 level-1
 isis hello-multiplier 4
 mpls ip
 mpls label protocol ldp
 ip pim sparse-mode
!
interface GigabitEthernet3/0/0
 description GigE to secondary local LAN
 ip address 195.111.101.210 255.255.255.0
 ip access-group accessfilter in
 service-policy output core-forwarding
 ip router isis
 isis circuit-type level-1
 isis metric 30 level-1
 isis hello-interval 10 level-1
 isis hello-multiplier 4
 mpls ip
 mpls label protocol ldp
 ip pim sparse-mode
!
interface Ethernet 4/0/0
 description Link to Management Network
 ip address 10.0.0.210 255.255.255.0
 ip access-group mgmt-in in
 ip access-group mgmt-out out
!
interface lo0
```

```
 description Loopback
 ip address 195.111.255.210 255.255.255.255
 ip router isis
!
ip access-list 1 permit 195.111.255.0 0.0.0.255
ip access-list 99 permit host 195.111.253.5
ip access-list 99 permit host 195.111.253.6
ip access-list extended mgmt-in
 permit 10.0.0.0 0.0.0.255 host 195.111.255.1
 permit 10.0.0.0 0.0.0.255 host 10.0.0.1
 deny any any
ip access-list extended mgmt-out
 permit host 195.111.255.1 10.0.0.0 0.255.255.255
 permit host 10.0.0.1 10.0.0.0 0.255.255.255
 deny any any
ip access-list extended accessfilter
 permit ip 195.111.250.0 0.0.0.255 host 195.111.255.1 eq ssh
 permit ip 195.111.250.0 0.0.0.255 195.111.0.0 0.0.3.255 eq ssh
 permit ip 195.111.255.0 0.0.0.255 host 195.111.255.1 eq ssh
 permit ip 195.111.255.0 0.0.0.255 195.111.0.0 0.0.3.255 eq ssh
 permit ip 10.24.0.0 0.0.0.255 host 195.111.255.1 eq ssh
 permit ip 10.24.0.0 0.0.0.255 195.111.0.0 0.0.3.255 eq ssh
 deny ip any host 195.111.255.1 eq ssh
 deny ip any 195.111.0.0 0.0.3.255 eq ssh
 permit ip 195.111.250.0 0.0.0.255 host 195.111.255.1 eq ftp
 permit ip 195.111.250.0 0.0.0.255 host 195.111.255.1 eq ftp-data
 permit ip 195.111.255.0 0.0.0.255 host 195.111.255.1 eq ftp
 permit ip 195.111.255.0 0.0.0.255 host 195.111.255.1 eq ftp-data
 permit ip 195.111.250.0 0.0.0.255 195.111.0.0 0.0.3.255 eq ftp
 permit ip 195.111.250.0 0.0.0.255 195.111.0.0 0.0.3.255 eq ftp-data
 permit ip 195.111.255.0 0.0.0.255 195.111.0.0 0.0.3.255 eq ftp
 permit ip 195.111.255.0 0.0.0.255 195.111.0.0 0.0.3.255 eq ftp-data
 deny ip any host 195.111.255.1 eq ftp
 deny ip any host 195.111.255.1 eq ftp-data
 deny ip any 195.111.0.0 0.0.3.255 eq ftp
 deny ip any 195.111.0.0 0.0.3.255 eq ftp
 permit ip 195.111.250.0 0.0.0.255 host 195.111.255.1 eq telnet
 permit ip 195.111.250.0 0.0.0.255 195.111.0.0 0.0.3.255 eq telnet
 permit ip 195.111.255.0 0.0.0.255 host 195.111.255.1 eq telnet
 permit ip 195.111.255.0 0.0.0.255 195.111.0.0 0.0.3.255 eq telnet
 deny ip any host 195.111.255.1 eq ssh
 deny ip any 195.111.0.0 0.0.3.255 eq ssh
 permit ip any any
!
ip prefix-list my-martians deny 0.0.0.0/7
ip prefix-list my-martians deny 2.0.0.0/8
ip prefix-list my-martians deny 5.0.0.0/8
ip prefix-list my-martians deny 7.0.0.0/8
ip prefix-list my-martians deny 10.0.0.0/8
ip prefix-list my-martians deny 23.0.0.0/8
ip prefix-list my-martians deny 27.0.0.0/8
ip prefix-list my-martians deny 31.0.0.0/8
ip prefix-list my-martians deny 36.0.0.0/7
ip prefix-list my-martians deny 39.0.0.0/8
```

```
ip prefix-list my-martians deny 41.0.0.0/8
ip prefix-list my-martians deny 42.0.0.0/8
ip prefix-list my-martians deny 49.0.0.0/8
ip prefix-list my-martians deny 50.0.0.0/8
ip prefix-list my-martians deny 58.0.0.0/7
ip prefix-list my-martians deny 70.0.0.0/7
ip prefix-list my-martians deny 72.0.0.0/5
ip prefix-list my-martians deny 83.0.0.0/8
ip prefix-list my-martians deny 84.0.0.0/6
ip prefix-list my-martians deny 88.0.0.0/5
ip prefix-list my-martians deny 96.0.0.0/3
ip prefix-list my-martians deny 169.254.0.0/16
ip prefix-list my-martians deny 172.16.0.0/12
ip prefix-list my-martians deny 173.0.0.0/8
ip prefix-list my-martians deny 174.0.0.0/7
ip prefix-list my-martians deny 176.0.0.0/5
ip prefix-list my-martians deny 184.0.0.0/6
ip prefix-list my-martians deny 189.0.0.0/8
ip prefix-list my-martians deny 190.0.0.0/8
ip prefix-list my-martians deny 192.0.2.0/24
ip prefix-list my-martians deny 192.168.0.0/16
ip prefix-list my-martians deny 197.0.0.0/8
ip prefix-list my-martians deny 198.18.0.0/15
ip prefix-list my-martians deny 223.0.0.0/8
ip prefix-list my-martians deny 224.0.0.0/3
ip prefix-list my-martians permit 0.0.0.0/0 ge 8 le 28
!
ip community-list standard Customer permit 1:1
ip community-list standard Peer permit 1:2
ip community-list standard Transit permit 1:3
ip community-list standard DontLeakThis permit 1:999
ip community-list standard PleaseLeakThis permit 1:1000
ip community-list standard OurAggregates permit 1:50000
!
class-map match-any expedited-forwarding
 match ip precedence 101 110 111
 match mpls experimental 101 110 111
 match ip dscp ef cs5 cs6 cs7
class-map match-any assured-forwarding
 match ip precedence 011
 match mpls experimental 011
 match ip dscp af31 af32 af33
class-map match-any best-effort
 match ip precedence 000 010 100
 match mpls experimental 000 010 100
 match ip dscp default cs2 cs4
class-map match-any less-than-best-effort
 match ip precedence 001
 match mpls experimental 001
 match ip dscp cs1
!
policy-map core-forwarding
 class expedited-forwarding
 bandwidth percent 10
```

```
 class assured-forwarding
 bandwidth percent 10
 class best-effort
 bandwidth percent 79
 class less-than-best-effort
 bandwidth percent 1
!
router isis
 net 49.0000.1921.6825.5210.00
 is-type level-1-2
!
router bgp 1
 no synchronize
 neighbor LONDON-RR-CLUSTER peer-group
 neighbor LONDON-RR-CLUSTER remote-as 1
 neighbor LONDON-RR-CLUSTER send-community
 neighbor LONDON-RR-CLUSTER update-source loopback0
 neighbor LONDON-RR-CLUSTER next-hop-self
 neighbor TRANSIT-PROVIDER peer-group
 neighbor TRANSIT-PROVIDER send-community
 neighbor TRANSIT-PROVIDER prefix-list my-martians in
 neighbor TRANSIT-PROVIDER prefix-list my-martians out
 neighbor TRANSIT-PROVIDER route-map transit-in in
 neighbor TRANSIT-PROVIDER route-map transit-out out
 neighbor 195.111.255.1 peer-group LONDON-RR-CLUSTER
 neighbor 195.111.255.2 peer-group LONDON-RR-CLUSTER
 neighbor 158.175.0.2 peer-group TRANSIT-PROVIDER
 neighbor 158.175.0.2 remote-as 2001
 neighbor 158.175.0.6 peer-group TRANSIT-PROVIDER
 neighbor 158.175.0.6 remote-as 3002
 no auto-summary
!
ip bgp-community new-format
!
route-map transit-in permit 10
 set community Transit additive
!
route-map transit-out permit 10
 match community PleaseLeakThis
!
route-map transit-out permit 20
 match community Customer
!
route-map transit-out permit 30
 match community OurAggregates
!
ip pim rp-address 195.111.255.254
!
aaa group server my-radius-group
  server host 195.111.253.2 auth-port 1812 acct-port 1813 key <string>
  server host 195.111.254.2 auth-port 1812 acct-port 1813 key <string>
!
snmp-server community <string> ro 99
snmp-server host 195.111.253.5 traps version 2c <string>
```

```
snmp-server host 195.111.253.6 traps version 2c <string>
snmp-server enable traps
snmp-server trap-source loopback0
snmp-server contact Netops +44 1999 999999
snmp-server location London
!
ntp bootserver 195.111.253.4
ntp server 195.111.253.4
ntp server 195.111.254.4
!
no ip http server
no ip http server-secure
!
login banner ^C
            +---------------------------------------+
            |                                       |
            | ********** W A R N I N G ********** |
            |                                       |
            |    STRICTLY NO UNAUTHORISED ACCESS    |
            |                                       |
            +---------------------------------------+
^C
!
line con 0
 login authentication use-radius
 exec-timeout 0 0
 transport preferred none
line vty 0 4
 login authentication use-radius
 exec-timeout 0 0
 transport preferred none
```

References

Doyle, J. *Routing TCP/IP*, vol. 1, Cisco Press.

Doyle, J. and DeHaven Carroll, J. *Routing TCP/IP*, vol. 2, Cisco Press.

Doyle, J. and Kolon, M. (eds) *Complete Reference: Juniper Networks Router*, Osborne McGraw-Hill.

REQUESTS FOR COMMENTS

RFC 1825: Atkinson, R.—Security Architecture for the Internet Protocol—August 1995.

RFC 1918: Rekhter, Y. *et al.*—Address Allocation for Private Internets—February 1996.

RFC 2271: Harrington, D. *et al.*—An Architecture for Describing SNMP Management Frameworks—January 1998.

RFC 2272: Case, J. *et al.*—Message Processing and Dispatching for the Simple Network Management Protocol (SNMP)—January 1998.

RFC 2273: Levi D. *et al.*—SNMPv3 Applications—January 1998.

RFC 2274: Blumenthal, U. and Wijnen, B.—User-based Security Model (USM) for version 3 of the Simple Network Management Protocol (SNMPv3)—January 1998.

RFC 2275: Wijnen, B. *et al.*—View-based Access Control Model (VACM) for the Simple Network Management Protocol (SNMP)—January 1998.

RFC 2328: Moy, J.—OSPF Version 2—April 1998.

RFC 2375: Hinden, R. and Deering, S.—IPv6 Multicast Address Assignments—July 1998.

RFC 2401: Kent, S and Atkinson, R.—Security Architecture for the Internet Protocol—November 1998.

Designing and Developing Scalable IP Networks   G. Davies
© 2004 Guy Davies   ISBN: 0-470-86739-6

RFC 2460: Deering, S. and Hinden, R.—Internet Protocol, Version 6 (IPv6) Specification—December 1998.

RFC 2615: Malis, A. and Simpson, W.—PPP over SONET/SDH—June 1999.

RFC 2661: Townsley, W. *et al.*—Layer Two Tunneling Protocol 'L2TP'—August 1999.

RFC 2892: Tsiang, D. and Suwala, G.—The Cisco SRP MAC Layer Protocol—August 2000.

RFC 3345: McPherson, D. *et al.* Border Gateway Protocol (BGP) Persistent Route Oscillation Condition—August 2002.

RFC 3473: Berger, L. (ed.)—Generalized Multi-Protocol Label Switching (GMPLS) Signaling Resource ReserVation Protocol-Traffic Engineering (RSVP-TE) Extensions—January 2003.

RFC 3478: Leelanivas, M. *et al.*—Graceful Restart Mechanism for Label Distribution Protocol—February 2003.

RFC 3513: Hinden, R. and Deering, S.—Internet Protocol Version 6 (IPv6) Addressing Architecture—April 2003.

RFC 3588: Calhoun, P. *et al.*—Diameter Base Protocol—September 2003.

RIPE RECOMMENDATIONS

RIPE-229: Panigl, C., Schmitz, J., Smith, P. and Vistoli, C.—RIPE Routing-WG Recommendations for Coordinated Route-Klap Damping Parameters.

IETF INTERNET DRAFTS

draft-ietf-13vpn-rfc2547bis-01.txt—Rosen, E. and Rekhter, Y.—BGP/MPLS IP VPNs—Work In Progress.

draft-ietf-idr-bgp-ext-communities-06.txt—Sangli, S. *et al.*—BGP Extended Communities Attribute— Work In Progress.

draft-martini-12circuit-trans-mpls-13.txt—Martini, L. *et al.*—Transport of Layer 2 Frames Over MPLS—Work In Progress.

draft-martini-12circuit-encap-mpls-06.txt—Martini, L. *et al.*—Encapsulation Methods for Transport of Layer 2 Frames Over MPLS—Work In Progress.

draft-kompella-12vpn-12vpn-00.txt—Kompella, K. (ed.)—Layer 2 VPNs Over Tunnels—Work In Progress.

draft-ietf-13vpn-bgpvpn-auto-01.txt—Ould-Brahim, H., Rosen, E. and Rekhter, Y.(eds)—Using BGP as an Auto-Discovery Mechanism for Provider-provisioned VPNs—Work In Progress.

draft-ietf-12vpn-radius-pe-discovery-05.txt—Heinanen, J. and Weber, G. (ed.)—Using RADIUS for PE-Based VPN Discovery—Work In Progress.

Index

Designing and Developing Scalable IP Networks   G. Davies
© 2004 Guy Davies   ISBN: 0-470-86739-6